Techniques in Life Science and Biomedicine for the Non-Expert

Series Editor
Alexander E. Kalyuzhny, University of Minnesota
Minneapolis, MN, USA

The goal of this series is to provide concise but thorough introductory guides to various scientific techniques, aimed at both the non-expert researcher and novice scientist. Each book will highlight the advantages and limitations of the technique being covered, identify the experiments to which the technique is best suited, and include numerous figures to help better illustrate and explain the technique to the reader. Currently, there is an abundance of books and journals offering various scientific techniques to experts, but these resources, written in technical scientific jargon, can be difficult for the non-expert, whether an experienced scientist from a different discipline or a new researcher, to understand and follow. These techniques, however, may in fact be quite useful to the non-expert due to the interdisciplinary nature of numerous disciplines, and the lack of sufficient comprehensible guides to such techniques can and does slow down research and lead to employing inadequate techniques, resulting in inaccurate data. This series sets out to fill the gap in this much needed scientific resource.

More information about this series at https://link.springer.com/bookseries/13601

Shanmugasundaram Ganapathy-Kanniappan

Isotopic Tracer Techniques in Preclinical Research

 Springer

Shanmugasundaram Ganapathy-Kanniappan
Division of Interventional Radiology
Russell H. Morgan Department of Radiology & Radiological Science
Johns Hopkins University School of Medicine
Baltimore, MD, USA

ISSN 2367-1114 ISSN 2367-1122 (electronic)
Techniques in Life Science and Biomedicine for the Non-Expert
ISBN 978-3-030-99702-1 ISBN 978-3-030-99700-7 (eBook)
https://doi.org/10.1007/978-3-030-99700-7

This Springer imprint is published by the registered company Springer Nature Switzerland AG
The registered company address is: Gewerbestrasse 11, 6330 Cham, Switzerland

To my parents, grandparents, brother, sister-in-law, and nieces, to whom I owe everything.

Preface

The propensity to understand processes or events is an inherent nature of mankind. This natural curiosity to follow or trace objects in the external as well as internal milieu to gain knowledge is the foundation of science. Hence, the conceptual history of the idea of "tracing" is as ancient as human evolution. As aptly quoted by Robertson, "The origins of tracer concept are lost on antiquity if this is taken to include such examples as the use of the cowbell to locate a herd and similar applications…" Understandably, tracers include a vast array of agents of diverse chemical nature. The primary focus of this book is on the techniques related to isotopic tracers in preclinical research.

The dawn of the twentieth century witnessed remarkable progress in the field of nuclear science that nurtured the emergence of "nuclear medicine," one of the sophisticated fields of modern medicine. Over 100 years ago, the first unstable (radioactive) isotope was discovered while extracting the element, lead (Pb) from radium. This seminal discovery by the Hungarian-Swedish scientist, Georg Charles von Hevesy (1911) documented the occurrence of radioisotope ^{210}Pb, thus marking the beginning of the era of isotopic tracers. With the advances in the synthesis and application of novel isotopes concurrent with the ingenuity of medical physics, the field of diagnostic imaging and clinical management expanded to new heights. From the basic science perspective, the advent of isotopic tracers and their utilization in research led to significant improvement in our ability to decipher molecular interactions at macro- and microscopic levels and elucidation of vital physiologic mechanisms (e.g. metabolic regulation). In other words, isotopic tracers provided a new armamentarium to push the limits of investigation in fundamental science, preclinical as well as translational research.

The primary objective of this book, in line with the thematic series "Techniques for Nonexperts," is to provide a detailed overview of the isotopic tracer techniques currently employed in experimental biology as well as in clinically relevant disciplines such as experimental therapeutics and translational research. The book is intended to serve as a practical guide and handbook on the application and analysis of isotopic tracers in research. Accordingly, the techniques discussed and outlined were carefully chosen based on their relevance and significance in the analysis of cellular and molecular biochemistry, as well as human pathophysiology.

To familiarize the reader with the tracer techniques, each protocol begins with a brief background, principles, detailed methodology including reagents required, and stepwise experimental analysis. The author hopes that this handbook will be a treasured research manual of tracer techniques. This book will be valuable to students, researchers as well as teachers interested in the application of isotopic tracers in experimental and diagnostic research.

I wish to thank all the researchers and scientists whose contributions advanced the understanding and implications of isotopic tracers in nuclear physics, pharmaceuticals, and nuclear medicine. I also thank all the researchers/publishers for their kind consent to reproduce some of the illustrations which will enable an easy and effective understanding of the protocols. The Springer publishers' constant support and encouragement made this book a reality.

Baltimore, MD, USA   Shanmugasundaram Ganapathy-Kanniappan

Acknowledgement

This book pays homage to the early great chemists, physicists, biochemists, and biologists who laid the foundation for the modern field of nuclear medicine that encompasses radioactive and nonradioactive tracers. The impact of tracer techniques is manifold as exemplified by their role in molecular biology, proteomics, metabolomics, radiopharmaceuticals, nuclear medicine, and clinical imaging. Current knowledge in the application of tracers, especially the isotopic labeling in science and medicine is a result of the enormous contributions of countless researchers of the past and present. I take this opportunity to gratefully acknowledge the efforts and inputs of all, including but not limited to the students, research scholars, postdoctoral fellows, medical physicists, nuclear physicists, clinicians, clinician-scientists, nurses, and technicians who helped in the advancement of the field of isotopic tracers. Last but not the least, with profound gratitude I thank the Springer publishers, especially the team associated with this project for their unflinching kindness, patience and constant support during this difficult pandemic period.

Contents

About the Author

Shanmugasundaram Ganapathy-Kanniappan holds a Ph.D. degree from the University of Madras, India. In 2006, he joined the Department of Radiology, Johns Hopkins University School of Medicine as a junior faculty member and was quickly promoted to assistant professor. Before joining the Johns Hopkins, he underwent postdoctoral training in premier institutes such as the National Institute of Immunology (NII, New Delhi, India) and the University of California at Los Angeles (UCLA). His current research interests include metabolic alteration in human pathophysiology and the identification of molecular targets for therapeutic intervention. He has published over 50 articles including original reports, invited reviews, editorials, book chapters, and books (https://www.ncbi.nlm.nih.gov/myncbi/1xKsymd-il85x/bibliography/public/). He is an inventor and coinventor of several patents and a recipient of prestigious fellowships and research awards including funding from private and federal sponsors. He is on the editorial board of peer-reviewed journals and currently serves the journal, Frontiers in Oncology as an associate editor for the section, Cancer Molecular Targets and Therapeutics.

Chapter 1
Introduction

1.1 Historical Background of Isotopes

A tracer is any substance or system that enables the tracking of a molecule or pathway involved in chemical, biological, or physical processes. The earliest indication that tracers can be used to investigate mechanisms of biological phenomenon dates back to the early eighteenth century. The concept of tracing in biology was introduced in 1736 by John Belchier, an English surgeon who demonstrated that the bones of animals that fed upon *madder roots* had distinctive color (a deep red dye) but not in other animals. Belchier's seminal discovery not only suggested the feasibility of tracing but provided the first clue that nutrient metabolism may be trackable. With tremendous progress in nuclear chemistry including radiochemistry and synthetic chemistry, the development of novel tracers as powerful tools has reached significant heights. Particularly, their relevance in human health management (e.g., diagnosis) has necessitated the emergence of the field of nuclear medicine. A vast literature is available on the chemistry and the fundamental principles underlying the synthesis of tracers. Similarly, substantial literature demonstrates the application of tracers in various phenotypes and organ-specific investigations. In this book, through an eclectic approach, we intend to provide a detailed guideline on the methods and/or technical application of tracers as relevant to experimental research. Next, the term tracer refers to a broad range of chemicals or agents that has the utility in tracking any pathway or process. For example, reporter tags such as fluorescent probes, coloring dyes, and isotopes among others are frequently employed in tracer studies. The emphasis here will be on the current and recent trends in isotopic tracer studies as applied to in vitro and in vivo models representing basic, preclinical, or translational research.

In the interest of the readers, before going into the details on the type of techniques, etc., it may be useful to introduce the basic definition of isotope, stable isotope, and radioactive isotope, the terms that are frequently used throughout the

© Springer Nature Switzerland AG 2022
S. Ganapathy-Kanniappan, *Isotopic Tracer Techniques in Preclinical Research*,
Techniques in Life Science and Biomedicine for the Non-Expert,
https://doi.org/10.1007/978-3-030-99700-7_1

book. In the word, isotope, the prefix "iso" has its origin from the Greek word "isos" meaning similar or equal, whereas the "tope" is derived from the Greek word "topos" referring to the location or position. As we know, the chemical elements are segregated based on their atomic number (i.e., the total number of protons) and arranged in the Periodic Table. An isotope of an element (e.g., Hydrogen, H) also has the same atomic number as the naturally abundant H. However, the isotope differs in the atomic mass (i.e., the sum of protons and neutrons). Thus, in the case of H, the naturally abundant form of the element has 1 proton and no (0) neutron (i.e., 1 + 0), hence denoted as ^{1}H. However, two other forms, i.e., isotopes of H are also known that exhibit either 1 neutron or 2 neutrons in addition to the 1 proton. In other words, the isotopic forms of H have an atomic mass of 2 and 3, and are represented as ^{2}H (Deuterium) and ^{3}H (Tritium), respectively. Noteworthy, the Periodic Table which uses the atomic number (not the mass) to group the elements, thus shows all the three forms of H in position 1. In other words, one or more species of the same element that exhibit similar or equal ("isos") atomic numbers are located ("topos") in the same group on the Period Table.

1.2   Types of Isotopes and Tracer Techniques

Isotopes are broadly categorized as "stable and unstable" based on their chemical property to undergo spontaneous disintegration. Stable isotopes don't undergo decay or disintegration, whereas the unstable isotopes emit high-energy particles or radiation, hence referred to as radioactive isotopes or radioisotopes. As pointed out in the case of H, among the two isotopes, deuterium (^{2}H) is stable, whereas tritium (^{3}H) is unstable (radioisotope). Thus, the term isotope may indicate either stable or radioactive isotope unless and otherwise specified (e.g., radioisotope).

1.2.1   Radioactive Isotopes

In biological as well as clinically relevant analysis, isotopes that emit γ or β radiations are widely used. Quantification of beta (β)-radiation and gamma (γ)-radiation are performed by Geiger Muller (GM) counter, autoradiography, and liquid scintillation counters (for γ rays). Some of the most commonly employed isotopes in biological and/or clinical investigation with their type of radiation (β or γ) are ^{18}F-γ, ^{3}H-β, ^{11}C-γ, ^{14}C-β, ^{32}P, ^{33}P, and ^{35}S-β. Radioactive tracers are highly specific and distinctive that as low as picogram (pg) quantities, i.e., up to 10^{-12} units, may be detected. Next, in general, the radioactive tracers remain unaffected by conditions of the biological system such as temperature, physical barriers (muscle, bone), and other factors. Thus, the radioactive isotope signals, or data on the overall intensity of the tracer, or its distribution or specific location are reliable and reproducible. However, one of the challenges of radioactive tracers is their safe handling due to the radiation effects and also the need for a dedicated, licensed, and authorized facility for the synthesis and/or use of radioactive nuclides.

The choice of using a particular radioactive tracer in a biological system depends upon the half-life of the tracer, type of radiation emitted by the tracer, the facility to measure the specific radiation, and most importantly the compliance to radiation safety and legal/local regulations of the investigator/institute. The half-life of the radioactive tracer is one of the principal limiting factors. For example, the clinically used ^{18}F (fluorine isotope) has a half-life of 109 minutes which is less than 2 h, i.e., the first half of decay occurs in <2 h. In other words, from the time ^{18}F is synthesized, it has to be transported, employed in the clinical investigation, and imaging data acquired within 2 h to avoid its first half decay. On the other hand, other isotopes are very short-lived (e.g., nitrogen (^{13}N) and oxygen (^{15}O)) and require dedicated facilities with cyclotrons or accelerators to carry out the studies at a swift and fast pace.

Data collection or the endpoint in radiotracer studies is usually quantitative or qualitative imaging of specific radiation. For instance, γ-emitters in a sample may be quantified using a γ-counter, while β emitters may be measured using scintillation fluids in a β-counter. The qualitative analysis, at least in the experimental research, may be performed by capturing the image of radiation in x-rays films. In this context, it is helpful to understand the film-based image capturing techniques, namely, autoradiography and fluorography. In autoradiography, the image is captured by direct film exposure to the radioactive source (i.e.) radioisotope containing materials such as gels, membranes, etc., whereas in fluorography, the radioactive signal is enhanced by a secondary agent such as fluor or intensifying screens prior to the image acquisition by film exposure.

Radioactive tracers have been in use to assess cardiovascular function in humans for over 90 years. As early as 1927, Blumgart and their team documented the utility of radiotracers in measuring pulmonary circulation time. Radioisotopes have been proven to be valuable in other emerging technologies such as metabolic imaging of tumors using radiolabeled amino acids (e.g., ^{11}C-methionine, ^{11}C-tyrosine). Radioisotopes are also used in the characterization and sequencing of oligosaccharides of glycoproteins using radiolabeled sugars or carbohydrates. With the advances in detection techniques and the type of radiopharmaceuticals, the application of radiotracers has increased tremendously.

1.2.2   Stable Isotopes

Unlike the radioactive isotopes, in stable isotopes, the absence of hazardous radiation renders them a unique tool to be used in humans and animals. The first evidence on the utility of stable isotopes in clinical diagnosis/research predates to the 1930s. With the availability of a variety of stable isotopes, their application encompasses wide ranges of research including the less-known macronutrients and mineral metabolism. In macronutrients research, the commonly used stable isotopes include

^{2}H, ^{13}C, ^{15}N, and ^{18}O. Similarly, in mineral metabolism research, stable isotopes such as ^{25}Mg, ^{26}Mg, ^{42}Ca, ^{46}Ca, ^{48}Ca, ^{57}Fe, ^{58}Fe, ^{67}Zn, and ^{70}Zn are frequently used. Besides these, in preclinical, translational investigations and human metabolism research [1-^{13}C] leucine and [1-^{13}C, ^{15}N] leucine are widely used as tracers. Besides the in vitro (cellular) studies stable isotopes have also been employed in vivo to investigate metabolism in animals as well as humans.

Stable (nonradioactive) isotopes have a major impact on metabolism research. The emergence of the application of stable isotopes has significantly advanced our understanding of the kinetics of metabolism, specifically the synthesis, depletion, or alteration in the turnover of various metabolites in diverse human pathophysiology. Tracer studies have advanced our understanding of Zinc metabolism and homeostasis, rolandic epilepsy, etc. The advent and improvisation of liquid-chromatography (LC) coupled with mass spectrometry (MS) have remarkably elevated the efficiency and sensitivity of tracer-dependent human research in metabolism.

Stable isotope-labeled metabolites have also been employed in the investigation of metabolism in preclinical tumor models (e.g., ^{13}C-glucose; ^{13}C, ^{15}N-glutamine). In fact, ^{13}C-leucine is one of the labeled amino acids of choice that is extensively used in the investigation of protein turnover at the whole-body level. Most recently, in a mouse model, it has also been demonstrated that a stress-free method of administration of such stable-isotope labeled metabolites is feasible. Using a liquid diet it is documented that stable isotopes may be delivered with efficiency and used for tracing metabolic networks at the organismal level.

Unlike the radioisotopes, the data collection in experiments of stable isotope tracers involves either chromatographic fractionation of isotope-labeled molecules followed by MS or magnetic resonance imaging (MRI) or magnetic resonance spectroscopy (MRS). The fractionation by high-performance liquid chromatography (HPLC) followed by ionization and identification of isotope-labeled molecules by tandem MS (MS/MS) is a widely employed approach for the quantitation as well as characterization of the molecule of interest. MRS and MRI involve advanced, sophisticated high efficient equipment that enables live imaging at cellular and organismal levels.

One of the significant advantages of stable isotopes is their safe use in clinical investigation besides preclinical research and basic science. Prominent features of stable isotopes that distinguish them from their radioactive counterparts are considered advantageous in clinical research. For instance, unlike the radioisotopes, it is possible and safe to use multiple stable isotopes concurrently in the same subject with no undesirable effects on other investigations or future studies. Next, the amount of sample required for the analysis of stable isotope is significantly small hence can be adopted for children and infants. Finally, as mentioned elsewhere, unlike the hazardous effects, handling, and disposal of radioactive wastes, the use of stable isotope is safe for both the participants/patients and staff/technicians involved in the study. Thus, the stable isotopes enable the real-time kinetic assessment of metabolites or metabolic pathways in disease conditions as well as in prognosis.

1.3   Overview of the General Applications of Isotopic Tracers in Basic and Preclinical Research

The techniques presented here are broadly grouped as relevant in functional research in basic science and applied science otherwise known as translational research. In the basic science segment, both radioactive and non-radioactive (stable) isotopes are covered, whereas, in the translation research, the preclinical drug development approaches such as pharmacokinetics and pharmacodynamics are explained using a specific therapeutic as a model.

The isotopic tracers irrespective of the stability (or decay) serve as powerful tools to investigate otherwise challenging research problems. Much of the knowledge on the transcriptional regulation of specific genes by nuclear and non-nuclear proteins is an outcome of extensive application of radioisotopes in DNA, RNA, and protein biology. Hence, the book begins with the role of specific radioactive isotopes as tracers in research involving DNA-DNA interaction (Southern blotting), RNA-DNA interaction (Northern blotting), and nucleic acid (e.g., DNA)-protein interaction (electrophoretic mobility shift assay). Furthermore, the famous pulse-chase experiment that demonstrated the ability to study protein synthesis using radioisotope labeling of amino acids is also included.

Next, the application of stable isotopes in protein analysis, especially in quantitative and qualitative proteomics, is discussed using two widely employed methods: SILAC (Stable Isotope Labeling by Amino acids in Cell culture) and iTRAQ (Isobaric Tag for Relative and Absolute Quantitation). The former is relevant in the analysis of protein profiles of two samples, and facilitate the tracing and mechanistic delineation of protein regulation in a cellular environment. The latter is a method of choice for the analysis of tissues or body fluids that are already available or extracted, and enables concomitant characterization of proteins from multiple samples.

In the next section, i.e., applied science or translational investigation, the majority of the approaches are exemplified with radioisotopes due to the preponderance of data and literature on the validation of such approaches. However, it must be noted that in the light of recent research, it may be possible to employ stable isotopes, if not in all cases, at least in some methodologies. Nevertheless, our experience and expertise in radioisotopes in identifying drug targets and evaluation of the kinetics/distribution of experimental therapeutics instigated the inclusion of the well-established and widely employed radiotracer techniques.

Chapter 2
Tracer Technique in Basic Research

2.1 Introduction

Two important and indispensable methods that are integrated into modern analytical biochemistry and molecular biology are electrophoresis, i.e., resolution of molecules such as proteins and nucleic acids, based on their charge and mass, and the ability to identify the specific target by "blotting." Thus, the development and improvements in gel electrophoresis and subsequent transfer onto a solid support system for further characterization of the specific identity have revolutionized the pace and depth of research in basic as well as applied science.

Until the development of modern nucleic acid analysis technologies (e.g., DNA sequencing), it was the resolving power of gel electrophoresis that was used to efficiently separate and profile nucleic acids based on the size and net charge. However, the determination of the specific identity of such resolved molecules remained a challenge. For instance, fragments of genomic DNA or plasmid DNA obtained by enzymic digestion may be separated on gel electrophoresis, stained with ethidium bromide followed by visualization of DNA fragments under ultraviolet light. However, the identity of a specific gene or nucleotide sequence could not be ascertained due to the lack of a probe. In this context, the hybridization technique (based on the principle of pairing of nucleic acids to complementary strands) enabled the use of a short fragment or nucleotide sequence to be used as a probe to verify and identify the specific gene. Yet, it involved the gel slicing and extraction of DNA fragments from the gel for further hybridization with probes. This process becomes further complicated if the study is about unknown or less-known genes. In such a scenario, all the fragments of a gel need to be isolated/extracted individually and hybridized.

Progress made in the mid-1970s on the method development to improve the analytical efficiency of nucleic acids and proteins greatly advanced the field of molecular biology. First and foremost, the demonstration by Dr. Edwin Southern on the

© Springer Nature Switzerland AG 2022
S. Ganapathy-Kanniappan, *Isotopic Tracer Techniques in Preclinical Research*,
Techniques in Life Science and Biomedicine for the Non-Expert,
https://doi.org/10.1007/978-3-030-99700-7_2

feasibility of transferring or blotting DNA from agarose gel onto a solid support membrane (e.g., cellulose nitrate) facilitated efficient molecular hybridization techniques to determine sequence identity using radioisotope labeled probes (e.g., ^{32}P-labeled nucleotide fragment). Furthermore, this method also laid the foundation for the development of subsequent protocols that are specific for RNA and protein, namely, Northern and Western blotting techniques, respectively.

2.2   Nucleic Acids Interaction: Southern Blotting

2.2.1   Background

Until the mid-1970s, specific characterization of biological molecules (e.g., DNA) separated by gel electrophoresis was a tedious and cumbersome process. It involved slicing (excise) of the specific spot or band of the gel (fragment), elution of the molecule (e.g., DNA) from the excised gel band followed by its hybridization with a known DNA or RNA sequence. In 1975, the pioneering work of Edwin Southern documented that nucleic (DNA) fragments resolved on an agarose gel can be transferred onto cellulose nitrate (nitrocellulose) membranes by simple capillary action. In this method, as the DNA from the gel is blotted onto the nitrocellulose membrane, this technique is referred to as a "blotting technique." In recognition of the significant contribution made by Edwin Southern, this blotting method is known as Southern blotting. Thus, Southern blotting is the first blotting technique introduced in molecular biology. This method formed the foundation for the subsequent advances in blotting techniques that became pivotal in the characterization of specificity or identification of various molecules such as RNA, protein, etc. It indeed revolutionized experimental molecular biology at least in the field of nucleic acids research including the protein-DNA interaction among others. Thus, Southern blotting is used to identify a specific DNA molecule or gene in a mixture of DNA fragments.

2.2.2   Method

The original method of Southern blotting has been improved and modified at several steps of the protocol; nevertheless, the principal components of the concept are retained and indispensable. The method described here is in line with the conventional protocol as adapted from the original report and widely employed in laboratories around the world. The basic requirement is the DNA sample to be analyzed which can be either genomic DNA from cells/tissues or a plasmid DNA or recombinant DNA. The DNA to be analyzed is enzymatically digested and fragmented followed by separation by gel electrophoresis. Following the electrophoretic resolution, the DNA fragments of the entire gel are blotted onto a nylon membrane through capillary transfer leading to the generation of an identical pattern of the DNA fragments as on the gel before the blotting. Once transferred, the nylon membrane containing the DNA fragments is hybridized with known probes, i.e., radiolabeled nucleotide sequence (^{32}P-labeled DNA), followed by visualization of the hybridization signal by autoradiography. Though the method is very simple and comprehensible, there are intricacies at various stages of the protocol that determine the efficiency and reliability of the final data, i.e., autoradiographic signal. Most importantly, care must be given to the preparation and handling of all reagents and chemicals used in the method to be free of nuclease contamination. Since DNA fragments are investigated, and the DNA probes are used for the detection, contamination of DNases or nucleases will severely compromise the integrity of the target as well as the probe. Next, careful blotting through capillary transfer, followed by incubation of DNA transferred onto the nylon membrane to denature the double-stranded DNA into single strands, is critical to facilitate probe hybridization. Similarly, the choice of probe selection, radiolabeling efficiency of the probe, and conditions of hybridization are all critical determinants of successful, specific probe-target hybridization. We will discuss these steps under the step-wise protocol.

2.2.3   Principle

The primary application of the Southern blotting technique is to determine/identify specific genes or DNA fragments. Southern blotting is based on the principle that capillary transfer of DNA fragments onto a solid support (e.g., nylon membrane) enables efficient hybridization and identification of specific genes using a radiolabeled probe. Although the resolution of DNA fragments on agarose gels and the probe-dependent hybridization methods were known before Southern blotting, the key step of blotting, i.e., the capillary transfer of DNA onto a membrane is the distinguishing feature of the latter that facilitates efficient and concurrent analysis of multiple samples.

2.2.4   Reagents/Equipment

Unless otherwise indicated all chemicals can be procured from Sigma Aldrich or similar suppliers.

- EDTA (ethylenediaminetetraacetic acid).
- 10X TAE buffer (Tris, 40 mM; acetic acid, 20 mM; EDTA 1 mM in double-distilled (dd), nuclease-free water (NFW); adjust the pH to 7.5–7.8).
- Agarose, analytical grade suitable for electrophoresis.
- DNA digesting enzymes (restriction enzymes with corresponding buffers).

- 6X-DNA gel loading dye (bromophenol blue, 0.25%; xylene cyanol FF, 0.25%; glycerol 30% in dd NFW).
- DNA molecular weight standard or markers.
- Ethidium bromide (EtBr) solution (0.5 µg/ml in dd NFW).
- Nylon membrane.
- Whatman 3MM filter paper.
- Stack of paper towels.
- SDS, sodium dodecyl sulfate.
- 20X Sodium chloride (NaCl) and sodium citrate ($Na_3C_6H_5O_7$) solution (SSC); (3 M NaCl and 0.3 M $Na_3C_6H_5O_7$ in dd NFW).
- 6X SSC solution (dilute the 20X SSC solution into 6X by dd NFW).
- 2X SSC solution (dilute the 20X SSC solution into 2X by dd NFW).
- 5X Denhardt's solution, (Ficoll, 0.02%; polyvinylpyrrolidone, 0.02%; bovine serum albumin (BSA), 0.02% in dd NFW).
- Hybridization solution (SDS, 0.5%; 6X SSC; 5X Denhardt's solution; sheared salmon sperm-DNA (100 mg/ml) in dd NFW).
- RNase A (at a concentration of 20 pg/ml in 2X SSC).
- Ultraviolet (UV) light source.
- Radioactive isotope (^{32}P).
- Radioisotope labeling kit/reagent (e.g., RadPrime DNA labeling system, Thermo Fisher Scientific).
- ^{32}P labeled-DNA probe.
- Gel electrophoresis equipment and power pack.
- Hybridization glass tubes.
- Hybridization oven.

2.2.5   Stepwise Protocol

The major steps of Southern blotting can be broadly divided into:

(i) DNA Restriction Digestion.

   The first step is to prepare the DNA sample for blotting. This necessitates the digestion of the DNA using restriction enzymes. The source of the DNA may be genomic DNA (i.e., isolated from cells or tissues) or plasmid DNA (i.e., cloned DNA or gene). Once the DNA is isolated from its source and purified/cleaned-up of excess salts or solvents (which may interfere with downstream reactions), the sample is ready for restriction enzyme digestion. The choice of restriction digestion is based on the literature or can be determined by the nature of the sample. For example, multiple restriction enzymes (e.g., EcoRI, KpnI) may be employed in genomic DNA digestion. In general, a 50 µl digestion reaction will include DNA sample (10 µg (10 µl)), 10X buffer corresponding to the restriction enzymes (5 µl), restriction enzymes (total of 5 µl (=50 units)), and dd NFW (30 µl). The reaction mix is incubated at 37 °C overnight (~16 h) followed by quenching and concentration to reduce the volume.

Note, it is also observed that addition of the enzymes may be in two stages, i.e., 30 units of enzymes added at the beginning followed by 20 units of enzymes halfway through the incubation period. For instance, in an overnight incubation, the first addition will be at the beginning, while the remaining enzymes will be added 2 h prior to the end of the reaction. In other incubation conditions such as 2 h of digestion (as in plasmid DNA samples), the 30 units of restriction enzymes will be added at the beginning, and an hour later the remaining 20 units of enzymes may be added to the reaction.

Next, the quenching of the restriction enzymes after digestion is performed by heat inactivation of the entire mixture at 65 °C for 20 min. The concentration of the digested DNA may be performed by ethanol precipitation which will also remove the salts and other chemicals of the digestion reaction. The ethanol precipitated DNA can be dissolved and made up in dd NFW for the subsequent gel electrophoresis. Noteworthy, the concentration of the entire digested mixture is optional but preferred, to reduce the total volume of the sample to facilitate loading onto the gel in the next step. In a 1 or 1.5 mm thickness agarose gel, commonly used combs can create a well to accommodate 25–40 µl volume; hence reducing the volume of the digested DNA to 20–30 µl is recommended (note, a gel loading dye (6×) of 4–6 µl will also be added to this sample before gel loading).

(ii) Agarose Gel Electrophoresis.

Using 1X TAE buffer agarose slurry is prepared to cast the gel for electrophoresis. Two essential factors to be considered are the percentage of the gel (i.e., porosity) and the handling of EtBr (for DNA visualization under UV light). In general, 1% gel is sufficient to resolve the DNA fragments of the restriction digestion. However, depending upon the size (kilobase, kb) of the fragment of interest, the gel percentage may be modified. For example, 1.2% agarose gel is sufficient to achieve better resolution or separation of fragments between 0.4 and 6 kb. On the other hand, 0.6% gel is good to separate DNA fragments of 1–20 kb. Next, the EtBr is mixed in the gel slurry before gel polymerization to enable UV-light visualization of the resolved DNA fragments following the electrophoresis. EtBr is carcinogenic, hence must be handled with utmost caution such as wearing gloves, goggles, and also appropriate disposal of the gels, buffers, and other materials exposed to EtBr.

Once, the agarose gel is cast, the digested DNA samples along with appropriate volume of gel loading dye (6×) are loaded onto the wells, and subjected to electrophoresis at a constant current. Depending upon the size (length and percentage) of the gel, the electrophoretic run may last for an hour or a few hours. Once the front running dye (bromophenol blue) reaches the other end of the gel, the electrophoresis is stopped. The gel needs to be carefully removed and visualized under UV and the image captured (for documentation purposes). It is always a good practice to place a ruler/scale next to the gel before

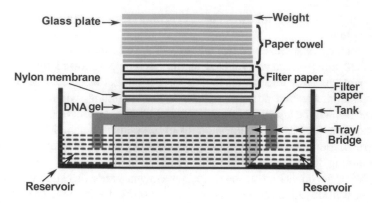

Fig. 2.1 Schematic showing the set-up for Southern blot transfer of DNA from agarose gel onto the nylon membrane. Due to the capillary action of the paper towels and filter papers staked on top of the nylon membrane, the buffer in the reservoir exhibits an upward migration/transfer. Consequently, the DNA fragments resolved in the gel also get transferred onto the nylon membrane. However, owing to the charge and pore-size of the membrane, the transferred DNA stays in the membrane while allowing the buffer to pass through to the staked paper towels. Note, the number of filter papers or the sheets of paper towels are shown as representation, and not actual or defined number which may be optimized as required by the research laboratory/investigator

capturing the gel image as it will provide us intricate details such as the orientation, distance of migration, position of fragments, etc.

(iii) DNA Transfer and Membrane Preparation for Hybridization

Prior to the transfer of DNA from the gel into the nylon membrane, the DNA fragments on the gel need to be denatured to facilitate subsequent probe hybridization. Hence, the gel with DNA fragments is immersed in a solution of NaCl (1.5 M) and NaOH (0.5 M) for 20–30 min, followed by neutralization of the gel with incubation in NaCl (1.5 M) and Tris-HCl (0.5 M, pH 7.5) for 30 min. Next, assemble the components of the transfer set-up as shown in Fig. 2.1. Care must be taken to arrange the gel, nylon membrane, and filter paper as well as paper towels without any air bubble between them, and also not overhanging to contact other stacked items (which may short circuit the capillary action). The duration of transfer depends upon the thickness of the gel, the size and amount of the DNA fragments to be transferred, etc. In general 12–16 h is sufficient for a complete transfer.

At the end of the transfer period, carefully remove the nylon membrane from the stack of filter papers and paper towels. Note, the gel may still be sticky and attached to the membrane, but it is normal. Next, using a pencil, mark the orientation of the gel on the nylon membrane. Next, carefully peel off the gel and discard, wash the membrane with 2X SSC for 5–10 min, and then crosslink the DNA onto the membrane by incubation at 80 °C for 2 h; alternatively, DNA crosslinking may be achieved by membrane exposure to short-wavelength UV light for 30 s. Note, commercial equipment (e.g., Stratalinker) is available for such UV crosslinking of DNA. Now the membrane is ready for prehybridization and hybridization with a probe.

The next step is to hybridize the DNA on the membrane with a specific probe, the ^{32}P-labeled nucleic acid. Before hybridization, to prevent non-specific binding of the probe to the membrane, a prehybridization or blocking step is recommended. This step uses the same solution that is employed in the hybridization except that it does not contain the probe. Place the membrane in a hybridization glass tube (vessel); care must be taken to avoid any air bubble between the membrane and the wall of the glass tube. Next, incubate the membrane in the prehybridization solution with sufficient volume to cover the entire membrane while in the hybridization oven for an hour. It is recommended to use the same stringent conditions that will be employed for hybridization. For example, if the hybridization is to be done at 40 or 65 °C, the prehybridization (blocking) needs to be performed at the same temperature. While the membrane is in blocking/prehybridization, prepare the probe for the subsequent step, i.e., hybridization.

(iv) Probe Synthesis

Probe refers to the short DNA or nucleotide sequence that is complementary or homologous to the target sequence of interest. In other words, the probe sequence has a base-pairing affinity to the specific gene or DNA fragment under investigation. Once the specific nucleotide sequence to be examined is confirmed, it can be synthesized or generated in multiple ways including PCR (polymerase chain reaction) amplification or nucleotide synthesis facility, etc. It is imperative that irrespective of the method of nucleotide synthesis, the resulting fragment must be purified to remove unused nucleotides and other chemicals used for the synthesis. The purified DNA thus generated is ready for labeling with ^{32}P, to obtain the "hot probe."

Labeling the DNA is one of the most critical steps that determine the accuracy of results. The efficiency of ^{32}P labeling of nucleotide fragment as well as the purification of the final labeled "hot probe" are critical. Caution: ^{32}P is a highly radioactive isotope; hence all safety measures of radiation must be strictly adopted in handling and disposal of the materials related to the entire protocol according to the institutional/organization's guidelines of the investigator. To generate an efficient "hot probe," several commercially available kits/reagents are available which the investigator may choose according to the resources and accessibility. Once the "hot probe" is synthesized, it is ready for hybridization.

(v) Hybridization

Prior to hybridization, the probe (DNA) needs to be denatured to generate single-strand DNA for pairing (hybridization) with the complementary target sequence of interest. Denature the probe DNA in an Eppendorf tube at 95 °C for 5 min, followed by immediate quenching in ice-bucket for 5 min. Then, briefly spin the tube to collect the contents at the bottom of the tube, and dilute it by adding a volume (e.g., 500 µl) of hybridization solution. Next, add the probe with the hybridization solution to the bottom of the glass tube containing the membrane. Care must be taken to avoid direct delivery of the probe onto the membrane. Continue the hybridization at the same temperature as prehybridization (e.g., 65 °C) at slow rotation speed. The duration of hybridization varies with the sample, probe, and other factors; ideally 4–5 h is sufficient unless the supplier of probe labeling kit or any hybridization solution indicates a different duration of hybridization.

(vi) Washing and Autoradiography

Following the hybridization, the membrane is carefully removed and washed twice with excess volume (e.g., ~200 ml) of 2X SSC (containing 0.1% SDS) for 15 min each, at room temperature. Next, for stringent washing, the membrane is washed twice (15 min each) with 0.2X SSC (with 0.1% SDS) at 65 °C. Meanwhile, carefully discard all the radioactive wastes including the probes according to the radiation safety guidelines of your institution/organization. Next, place the membrane on a Whatman 3MM filter paper, air-dry the membrane, and wrap it with Saran wrap to expose the membrane to an X-ray film. Follow the film exposure and developing procedures according to the supplier's instructions. The specific DNA-probe affinity/hybridization can be visualized as a strong autoradiographic signal.

2.2.6   Study Questions

What is the unique feature of Southern blotting that distinguishes it from previous methods?

2.3   Nucleic Acids Interaction: Northern Blotting

2.3.1   Background

The development of the blotting technique by Edwin Southern to study DNA sequences instigated the quest for a similar method to investigate gene expression (e.g., RNA). In 1977, Stanford scientists James Alwine, David Kemp, and George Stark demonstrated the feasibility of RNA analysis by blotting technique. It became known as the "Northern blot," with a sense of humor, and its analogy to Southern blotting.

The development of Northern blot expanded the analytical capacity of molecular biologists to investigate gene expression and characterization of specific RNA molecules. Northern blotting enables the determination of specific gene expression, quantification, and size characterization of particular transcript/gene identification of sequence specificity and/or alterations and so on. The repertoire of Northern blotting applications ranged from analysis genes in cell growth and development to the transformation of normal physiology into abnormal, patho-physiologic states of organs and tissues. With a wider application, the application of Northern blot expanded the depth of scientific knowledge in multiple subspecialties of molecular biology in the context of fundamental and application-oriented research.

2.3.2   Method

The fundamental feature of Northern blotting that distinguishes it from Southern blotting is the samples/targets to be analyzed are RNA molecules, not DNA. The concept and principle steps of the Northern blot are similar to the method of Southern blotting but with a focus on RNA. Consequently, the critical and most important technical challenge is the handling and maintenance of RNA in an RNase-free microenvironment throughout the experiment. Also, unlike DNA, the RNA is comparatively less-stable and prone to degradation in a myriad of conditions. Thus, RNA-care is a fundamental requirement for efficient Northern blotting.

The first and foremost step in this method is the sample preparation which is the isolation of RNA (maybe total RNA or messenger RNA (mRNA) from biological sources such as cells, tissues, fluids, etc. Once the RNA sample is ready for further analysis, a known quantity of RNA is subjected to agarose gel electrophoresis to resolve various species of RNA based on their size and charge. Following the electrophoretic resolution, the RNA from the gel is blotted onto a nylon membrane by capillary transfer. The transfer of RNA from the gel to a solid support system such as a nylon membrane enables probe hybridization and further analysis of specific genes of interest. In brief, the nylon membrane containing the RNA is hybridized with known probes, i.e., radiolabeled nucleotide sequence (^{32}P-labeled DNA), followed by visualization of the hybridization signal by autoradiography. Though the method is very direct and comprehensible, there are intricacies at various stages of the protocol that determine the efficiency and reliability of the final data, i.e., autoradiographic signal. Most importantly, care must be given to the preparation and handling of all reagents and chemicals used in the method to be free of RNase contamination. The detailed steps of the protocol will be discussed in the step-wise protocol.

2.3.3   Principle

Northern blotting technique encompasses several principles to achieve efficient quantification and analysis of RNA in a biological sample. First, it employs the principle that under denaturing electrophoretic conditions, RNA species of a sample may be resolved based on their size and net charge. Next, capillary transfer of the RNA thus resolved on a gel into a solid support enables further handling and analysis of different RNA species of a biological source. Finally, using radiolabeled complementary nucleotide sequence (DNA), it is possible to detect and quantitate a specific gene of interest and its correlation with physiology or pathophysiology.

2.3.4   Reagents/Equipment

- Unless otherwise indicated all chemicals can be procured from Sigma Aldrich or similar suppliers.
- EDTA (ethylenediaminetetraacetic acid).
- Ethidium bromide (EtBr) solution (0.5 µg/ml in RNase-free water).
- Diethylpyrocarbonate (DEPC).
- DEPC-treated water (DEPC 0.1% solution autoclaved for 20–30 min; stable at room temperature).
- RNase-free distilled water (molecular biology grade).
- 3-(N-morpholino)propanesulfonic acid (MOPS).
- 10× MOPS buffer (MOPS (0.2 M), sodium acetate (50 mM), and EDTA (1 mM), pH adjusted to 7.0 using NaOH).
- Agarose, analytical grade suitable for electrophoresis.
- Agarose gel [Agarose 1.2% in 1× MOPS buffer; formaldehyde (1%)].
- 2× RNA loading dye, [MOPS buffer (1×); glycerol (20%); formaldehyde (6.5%); formamide (50%); ethidium bromide (10 µg/ml); bromophenol blue (0.05%); xylene cyanol (0.05%)].
- RNA ladder (standards).
- Electrophoresis/running buffer (MOPS 1×; 7% formaldehyde).
- 5X Denhardt's solution, (Ficoll, 0.02%; polyvinylpyrrolidone, 0.02%; bovine serum albumin (BSA), 0.02% in RNase-free water).
- Nylon membrane.
- Whatman 3MM filter paper.
- Stack of paper towels.
- SDS, sodium dodecyl sulfate.
- 20X sodium chloride (NaCl) and sodium citrate ($Na_3C_6H_5O_7$) solution (SSC); (3 M NaCl and 0.3 M $Na_3C_6H_5O_7$ in RNase-free water).
- 1X SSPE [NaCl (150 mM), NaH_2PO_4 (10 mM), and EDTA (1 mM) in RNase-free water].
- 2 M sodium acetate in RNase-free water; adjust the pH to 4.0.
- Prehybridization buffer [formamide (50%); SDS (0.2%); SSC (5×), Denhardt's solution (5×), salmon sperm DNA (100 µg/ml), and sodium phosphate (50 mM, pH 6.5)].
- RediPrime II kit (GE Healthcare, Amersham Biosciences).
- ^{32}P-labeled dCTP (10 mCi/ml; Amersham Biosciences).
- Ultraviolet (UV) light source.
- Gel electrophoresis equipment and power pack.
- Hybridization glass tubes.
- Hybridization oven.

2.3.5   Stepwise Protocol

The three major sections of the Northern blot protocol are:

(i) RNA Isolation, and Denaturing Gel Electrophoresis

The method of isolation of total RNA from the biological source depends on the nature of the sample as well as the investigator's resources/experience. For example, Trizol© reagent is commonly and widely used to isolate total RNA from cells, tissues, and liquid samples. Alternatively, the investigator may use a lysis buffer followed by the organic solvent mixture (phenol: chloroform: isoamyl alcohol in the ratio of 25:24:1) and the related steps as documented for their study. Once the RNA sample is isolated, its concentration is determined using absorbance or OD (optical density) at 260 nm in a spectrophotometer. Now, the RNA is ready for electrophoretic resolution under denaturing conditions.

Prepare the agarose gel (1.2%) in MOPS with formaldehyde (1%) final concentration. Note, the agarose gel is prepared by melting the agarose in RNase-free water in a microwave; let the agarose slurry cool down before adding the formaldehyde and MOPS buffer. Also, the formaldehyde is toxic and must be handled with caution in a chemical/fume-hood. For example, a gel with 100 ml of agarose (1.2%) slurry may be prepared as follows: Dissolve 1.2 g of agarose in 80 ml RNase-free water, and boil in the microwave (few minutes); let it cool down to ~60 °C (do not let it overcool, as it may solidify). Next, add 10 ml of 10X MOPS and 3 ml of formaldehyde (the commonly available stock is ~37%). Finally, make the volume to 100 ml with RNase-free water, and pour the gel in the gel-cassette/tray with comb/jigs. Usually, in 30–40 min the gel will be polymerized and ready for use. Once the gel is ready, remove the comb, fill the tank with running buffer, and let the gel remain in the buffer until the RNA samples are prepared for loading. The RNA sample (~20–25 µg) is mixed with an equal volume of 2× RNA loading dye, mixed and denatured by heating at 75 °C, for 10 min in a heating block. The RNA thus denatured is then loaded onto the gel, and in one of the wells/lanes, a marker, i.e., RNA molecular weight standards, will be loaded as a reference (according to the supplier's instructions). The RNA is then resolved at 120–125 V for 2–3 h depending upon the length and thickness of the gel.

(ii) RNA Transfer

The second critical step is the transfer of RNA resolved on the agarose gel onto a nylon membrane. Prior to the transfer, as soon as the gel is removed from the electrophoretic apparatus, it needs to be stained with EtBr to visualize the RNA resolution (and documentation). In brief, the gel staining is done by immersion in EtBr solution (0.05%) for 15 min followed by destaining in either the running buffer or the RNase-free water for 15–30 min with gentle shaking. Following the destaining, RNA bands are visualized under UV light, and the image is captured for documentation. The visualization of RNA may reveal if the RNA integrity is good or degraded (smear) and also if the RNA of

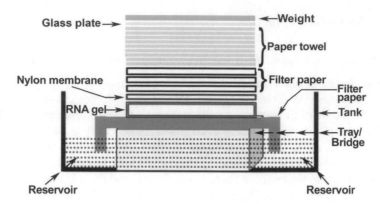

Fig. 2.2 Schematic showing the set-up for Northern blot transfer of RNA from agarose gel onto the nylon membrane. Due to the capillary action of the paper towels and filter papers staked on top of the nylon membrane, the buffer in the reservoir exhibits an upward migration/transfer. Consequently, the RNA fragments resolved in the gel also get transferred onto the nylon membrane. However, owing to the charge and pore-size of the membrane, the transferred RNA stays in the membrane while allowing the buffer to pass through to the staked paper towels. Note, the number of filter papers or the sheets of paper towels are shown as representation, and not actual or defined number which may be optimized as required by the research laboratory/investigator

each sample is loaded equally. Note, EtBr solution is carcinogenic, hence care must be taken for safe handling and disposal of EtBr as per institutional guidelines.

Before setting up the transfer, the gel is first immersed in 20X SSC for 15 min, and the nylon membrane in 2X SSC for 15 min. Then, the gel is assembled for blotting by capillary transfer as indicated for Southern blotting except for the specific differences in the sample as RNA, the buffer composition, RNase-free environment, etc. (Fig. 2.2). In brief, fill the buffer reservoir with 20X SSC; assemble a small tray in the middle of the reservoir to hold the sandwich of gel and other transfer materials. Place the Whatman 3MM filter paper soaked (in 20X SSC) on top of the tray with the ends of the filter paper dipping into the reservoir. Arrange three more pieces of Whatman filter paper soaked in 20X SSC over the existing Whatman paper on the tray. Next, place the gel followed by the wet nylon membrane; carefully remove any air bubble between the gel and the membrane. Arrange three pieces of Whatman filter paper soaked in 2X SSC over the nylon membrane. Next, at least for 3–4 inches, place the cut paper towels on top of the Whatman filter paper. On top of the paper towels place a weight (e.g., glass plate) to apply gentle pressure (of weight) to weigh down the whole transfer set-up. Transfer the RNA from the gel onto the nylon membrane overnight. After the transfer, carefully remove the weight, paper towels, and the filter paper. While removing the nylon membrane, carefully mark the orientation of the membrane with a pencil to indicate the gel-facing side as well as the lane of the RNA ladder/standards. Also, it is recommended to make a mark or cut one of the upper corners (e.g., left) of the membrane to identify the lanes of the membrane. Once the membrane is removed from the transfer set-up, crosslink the RNA onto the membrane by a UV Crosslinker (120 mJ/cm²) followed by incubation in a vacuum oven for 30 min. The membrane at this stage may be stored at room temperature if desired, otherwise, proceed with subsequent steps.

(iii) Hybridization and Film Developing

Similar to the Southern blotting, following transfer the nylon membrane is prehybridized meanwhile the DNA-probe complementary to the RNA species of interest is prepared for hybridization. Place the nylon membrane with the RNA side facing up, in the hybridization glass tube/vessel with 10–20 ml of prehybridization solution, and incubate at 68 or 42 °C for 1–2 h with gentle rotation in a hybridization oven. Meanwhile, prepare the probe by radioactive labeling of the DNA fragment (to be used to identify the RNA (gene) of interest). Using a commercially available kit (e.g., RediPrime II Kit) label 25–30 ng of DNA with [α-^{32}P] dCTP, and follow the instructions of the supplier to purify as well as evaluate the labeling efficiency. Before the addition of the probe, denature the DNA probe by heating in a boiling water bath for 5 min followed by snap cooling in a dry-ice bath. Add the probe to the hybridization solution, replace the prehybridization solution in the glass tube with the probe hybridization solution, and incubate the membrane at 68 or 42 °C overnight. Following the hybridization, remove the membrane and wash it with 1X SSPE containing 0.5% SDS at 65–68 °C for 15 min followed by two washes with 0.1X SSPE (containing 0.5% SDS) at 65–68 °C for 15 min each. Then, the membrane is removed, briefly air-dried on a blotting paper and covered with Saran wrap, and exposed to X-ray film for further development. In general, to improve the signal intensity the cassette with the film will be stored at −80 °C (for a few hours to a few days) until the film is processed/developed. A representative image of the agarose gel with RNA resolved by electrophoresis and Northern hybridization signal of a specific probe are shown Fig. 2.3.

2.3.6 Study Questions

Why is the RNA gel electrophoresis performed under denaturing condition?

2.4 DNA-Protein Interaction

2.4.1 Background

One of the applications of tracer technique that revolutionized the ability to characterize "protein interaction with nucleic acids" is the electrophoretic mobility shift assay (EMSA) also referred as "gel retardation assay." The EMSA technique has greatly advanced our understanding of gene regulation by proteins, particularly at

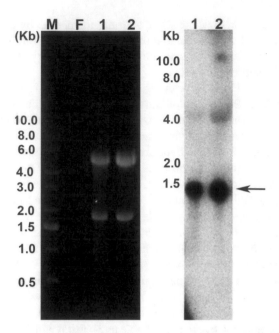

Fig. 2.3 RNA gel electrophoresis and Northern blotting of human GAPDH. *Left panel*, Ethidium bromide (EtBr) stained agarose gel showing the resolution of total RNA isolated from human hepatocellular carcinoma cell lines, Huh7 (1) and HepG2 (2). The two prominent gel bands illuminated by UV light correspond to the 18S and 28S ribosomal RNA (used as an indicator of RNA integrity as well as electrophoretic outcome). The RNA ladder (molecular marker, M) is also shown. *Right panel*, The RNA transferred and hybridized with a specific probe that correspond to human GAPDH (glyceraldehyde-3-phosphate dehydrogenase) showed a specific and strong signal of hybridization at 1.5 kb (indicated by arrow) in both Huh7 (1) and HepG2 cells (2). The lane F, left as free without any sample loading to avoid cross-contamination of the molecular markers. (Published from personal communication)

the transcription level whereby the binding of a protein(s) to a specific DNA sequence of a gene affects its expression in normal as well as pathophysiologic conditions. In other words, proteins that exhibit specific affinity and interaction with a nucleotide sequence regulate the upregulation or downregulation of the particular gene. This, DNA-protein interaction is pivotal in several cellular processes and has functional implications at the organ and organismic levels. In general, such proteins may be nuclear proteins (e.g., GATA) or cytoplasmic proteins (e.g., p53) with the capacity to translocate into the nucleus to carry out the specific function as may be necessary. Although primarily used in the qualitative analysis and functional evaluation, EMSA is also utilized in the quantitative analyses through kinetic parameters of either the DNA and/or the protein(s). Overall, EMSA is a powerful tool to characterize the DNA-protein molecular interactions and their function in a particular cellular phenotype. The same method can be adopted to investigate RNA-protein interaction by substituting an RNA probe instead of DNA. In this section, to avoid redundancy and maintain brevity, we discuss only the protocol pertinent to investigating DNA-protein interaction.

The earliest documentation of gel electrophoretic analysis of protein-nucleic acid complexes dates back to the early 1970s (Adams and Fried, 2007). A decade later, Garner and Revzin (1981) improvised the technique to analyze protein-DNA complexes, and little has changed since then. The application of EMSA from a molecular biology perspective is very profound and manifold. The advantages of EMSA in the DNA-protein interaction studies can reveal several functionalities of a protein(s) and also the corresponding nucleic acid sequence. Essentially, it is a principal technique to delineate the functions of "interacting systems" in biology. Some of the salient features of EMSA include; determination of the binding affinity of a protein to a specific nucleotide sequence; elucidation of the parameters that affect the binding kinetics or affinity of a protein to nucleotide sequence; evaluation of the affinities of various proteins to a specific nucleotide sequence and identification of protein complexes that bind to a specific nucleotide sequence to regulate gene expression. Finally, the source of protein (sample) to be analyzed could be cellular proteins (e.g., whole cell lysate, nuclear extract) or cell-free, synthesized proteins (e.g., in vitro transcription). Thus, EMSA enables the characterization of protein-DNA interaction and its relevance in particular cellular physiology.

2.4.2 Method

The principal method of EMSA involves the incubation of a protein or mixture of proteins obtained from cellular or tissue samples (or proteins synthesized extracellularly) with a radiolabeled-nucleotide fragment. The radioactive isotope, ^{32}P, is the most common and widely used isotope in EMSA. The nucleotide fragment refers to a short sequence of nucleic acid under investigation (subject of the study) to identify the protein(s) that bind with the target nucleotides. Once the short nucleotide fragment is labeled with ^{32}P, it is referred as a "hot probe" as we use it as a probe to identify the target protein(s) that has an affinity to the genetic sequence. The "hot probe" (^{32}P-labeled nucleotide fragment) is then incubated with the sample (proteins) and subjected to electrophoresis to separate the proteins that are bound or unbound with the "hot probe." Following the gel electrophoresis, the gel is dried and exposed to X-ray film for autoradiography. Based on the proteins resolved on the gel with the binding of the "hot probe," the X-ray film will demonstrate the radioactive signal. Several inferences can be ascertained based on the autoradiographic signal. For example, if the protein used for the investigation was known, then the autoradiography will confirm the binding of the protein with the nucleotide fragment. Next, if the protein of interest is in a sample mixture with other proteins (e.g., cell lysate), then additional experimental conditions like incubating the "hot probe" and the protein sample with and without the specific antibody (i.e., specific to the protein of interest) will yield a "supershift" in signal in autoradiography due to the increased molecular mass, i.e., a complex containing the "hot probe" + specific protein + the protein-specific antibody. This will validate the identity of the protein bound to the specific nucleotide sequence (hot probe).

2.4.3   Principle

EMSA is developed based on the principle that under non-denaturing, native electrophoretic conditions the mobility of protein bound-DNA will be slower compared to the free-DNA. In other words, the binding of a protein to the specific nucleotide fragment results in retardation of its mobility compared to the same nucleotide fragment that is free of any protein binding. Hence, the EMSA is also referred as "gel retardation assay." Next, labeling the nucleotide fragment with a radioactive isotope (e.g., ^{32}P) will result in the autoradiographic visualization of the free- as well as protein-bound nucleotide fragment (DNA), due to the difference in the mobility.

Furthermore, as the EMSA technique relies on the detection of the "hot probe," any addition of the same nucleotide fragment that is not labeled with the radioactive isotope will compete for the protein binding. Such, non-radioactive nucleotide fragment is referred as a "cold probe." The objective of this cold probe is to demonstrate the specificity of the protein-DNA binding. In other words, the visualization of a strong autoradiographic signal in the gel with "hot probe" will not be visible in a sample where the "cold probe" was also added to the reaction mixture containing the "hot probe." This principle is also used to test if a nucleotide fragment (other than the "hot probe") can also compete for the protein binding. For instance, if a nucleotide fragment A was used as the "hot probe," and it has a strong affinity to bind with a protein alpha (α), then its specific binding will be visible as a strong signal on the autoradiogram. However, if there is another nucleotide fragment, B (that differs in its affinity to bind with the same protein α, then the addition of fragment B ("cold probe") will not compete for binding, hence will not affect the strong signal generated by the "hot probe" and its binding with protein α. This type of competition is known as "non-specific competition," used as a negative control. On the contrary, if the nucleotide fragment B also has a binding affinity to the protein α, then its addition (as "cold probe") will result in the weak or absence of a radioactive signal on the autoradiogram due to its competition with the "hot probe." This type of competition with a "cold probe" that can also bind with the protein is referred as "specific competition." Thus, the EMSA is based on the electrophoretic mobility of the DNA-protein complex, and the ability to visualize the autoradiographic signal corresponding to the DNA-protein interaction.

2.4.4   Reagents/Equipment

Chemicals and reagents required for the preparation of gel polymerization for polyacrylamide gel electrophoresis (PAGE) include:

- Acrylamide mix (30%; 29:1 acrylamide-bisacrylamide).
- 10X Tris–acetate–EDTA (TAE) buffer (preparation: 400 mM Tris, 25 mM EDTA, and adjust the pH to 7.8 or 7.9 with acetic acid).
- Glycerol (20%).

- Nuclease-free (NF), double distilled (dd) water.
- Ammonium persulfate (10% solution).
- TEMED (N,N,N',N'-tetramethylethylenediamine).
- Nucleotide fragment (DNA) to be tested as the probe.
- ^{32}P-radioactive isotope (to be procured from authorized supplier as approved by investigator's institution/organization).
- ^{32}P labeling kit.
- Protein source (sample prepared from cell lysates or tissue homogenates or in vitro transcription, etc.).
- 10× Binding buffer (preparation: Tris (100 mM, pH 7.5), EDTA (10 mM), KCl (1 M), DTT (1 mM), glycerol (50%), BSA (Bovine serum albumin) (0.1 mg/ml)).
- 10X Dye/glycerol stock solution: Tris (10 mM), EDTA (1 mM), glycerol (50%), bromophenol blue (0.001%), xylene cyanol (0.001%) FF.
- Gel fixative solution: prepared by mixing methanol-acetic acid-water in the ratio of 2:1:7.
- Gel electrophoretic apparatus, and the electric power supply.
- Gel dryer apparatus.
- Autoradiography film cassette.
- X-ray films for autoradiography (e.g., GE healthcare, Kodak).
- Access to a darkroom equipped with film developer is needed for film autoradiography.
- Whatman filter paper 3MM (to support the gel during gel-drying).
- Saran wrap to cover the gel during exposure to the film.

2.4.5 Stepwise Protocol

(i) Sample Preparation

The source of the protein(s) to be investigated for its DNA binding activity is a critical factor in EMSA. First, the protein(s) needs to be in a solution/buffer that facilitates native, non-denatured conformation. Next, the sample containing the protein should be devoid of any potential interfering agents/chemicals that may hinder protein-DNA interaction. Several kits and sample preparation methods are commercially available to prepare proteins from cells or subcellular compartments (e.g., nucleus) or cell-free synthesis of peptides. Once the protein(s) sample is prepared and quantified using commonly available protocols such as BCA protein assay, etc., the sample is ready for downstream analysis.

(ii) Probe Labeling

Probe refers to the nucleotide fragment or the nucleotide sequence that is under investigation for its affinity to bind with the protein(s) of interest. Once the specific nucleotide sequence to be examined is identified, it can be synthesized or generated in multiple ways including PCR (polymerase chain reaction) amplification or nucleotide synthesis facility, etc. It is imperative that irrespective of the method of nucleotide synthesis, the resulting fragment must be purified to remove unused nucleotides and other chemicals used for the synthesis. The purified nucleotide fragment thus generated is ready for labeling with ^{32}P, to obtain the "hot probe."

Labeling the nucleotide fragment is one of the most critical steps of EMSA. The efficiency of ^{32}P labeling of the nucleotide fragment and the purification of the final labeled "hot probe" are critical. Caution: ^{32}P is a highly radioactive isotope; hence all safety measures of radiation must be strictly adopted in handling and disposal of the materials related to the entire EMSA according to the institutional/organization's guidelines of the investigator. To generate an efficiently labeled "hot probe," there are several commercially available kits/reagents available which the investigator may choose according to the resources and accessibility. Once the "hot probe" is synthesized, it is ready for the evaluation of its binding with the protein(s) of interest (i.e., DNA-protein interaction).

(iii) Gel Preparation and Pre-Run

Unlike the gel electrophoresis employed for western blotting and other applications, the EMSA employs native, non-denaturing PAGE. The gel preparation does not include any detergents or denaturing agents that would impair the DNA-protein interaction or affect the native conformation of proteins. Hence, gel preparation is performed with extreme caution, and all the solutions and reagents required are of analytical grade quality.

Also, gel polymerization is achieved by the mixing of monomers, acrylamide and bisacrylamide. There are many combinations of acrylamide-bisacrylamide such as 29:1, 37.5:1, and 70:1, and so on. Most of them work well unless the nature of the sample, i.e., the protein(s) and DNA complex, demands a particular mixture of monomers. However, the mobility of the DNA-protein complex depends on the percentage of the polymerization, which is represented as 5% or 6% or 7%, etc., and it is determined by the number of gel polymers present in the total volume of the gel casting solution. Thus, the ratio of the monomers (relies on acrylamide and bisacrylamide combination) and the percentage of the gel (amount of the polymer present in the total volume of the gel) are two different but essential aspects of PAGE.

If precast gels are available, the investigator is encouraged to utilize them as well.

In brief, prior to the preparation of gel-polymer, assemble the electrophoretic apparatus and set it aside to cast the gel. Next, the PAGE gel preparation of a 40 ml of polymer solution to cast a 4.5% gel is as follows in the given order:

Acrylamide mix (30%; 29:1 acrylamide-bisacrylamide) 6 ml.
10X TAE buffer 4 ml.
Glycerol (20%) 2 ml.
NF dd water 28 ml.
Ammonium persulfate (10% solution) 300 μl.
TEMED 30 μl.

Once the solutions are added, and mixed, pour the polymer solution into the assembled gel plates of the electrophoretic apparatus. Set aside for a few hours (1–2 h), and once the gel is polymerized it is ready for the electrophoretic run. Note, acrylamide is a neurotoxin; hence use proper gloves and protective measures to avoid direct contact or exposure.

It is always recommended that PAGE gel is subjected to a pre-run (i.e., without any sample loading) to initiate homogeneous current, solute mobility and also to verify if the gel density and conductance are homogeneous. Hence, after careful assembly of the gel into the apparatus, and proper addition of running buffers (1X TAE buffer), set the electrophoresis. It is also desirable to run the entire gel (pre-run) in the cold room or at 4 °C. Ideally, the pre-run will be slow at 10 volts per cm of the gel length. While the gel is in the pre-run, perform the next step, i.e., the binding reaction of the probe-protein.

(iv) DNA-Protein Binding Reaction

The binding reaction involves besides the "hot probe" and the "sample" (protein source) other components like Poly (dI: dC) (i.e., nucleotides for the inhibition of non-specific binging) and BSA (i.e., stabilizer of DNA-protein complex). A typical binding reaction assembled in a microcentrifuge tube (e.g., 0.5 or 1 ml Eppendorf tube) is shown below:

To 0.67 µl of the 10X-binding buffer, add 0.1 µg (1 µl) of the protein sample, 1 µl of hot probe (i.e., ^{32}P-labeled DNA fragment), Poly (dI: dC) 0.2 µl (1 µg/µl), BSA 1 µl (50 µg/µl), and NF dd water to make up the final volume of the reaction mixture to 10 µl. The volumes of each reactant can be scaled-up proportionately if required. Once the components are added, incubate the mixture at 30 °C for 30 min to 1 h. Note, the duration of incubation to achieve optimal binding reaction between DNA and the protein varies depending upon the nature of the sample and the DNA under investigation. Furthermore, the incubation of the reaction mixture in some cases is preferred at 4 °C or on ice. This is relevant in conditions where non-specific interaction of proteins with DNA (hot probe) is reported to occur at room temperature or 30 °C. Hence, the incubation conditions (such as the temperature, duration) need to be optimized by the investigator or adopted from prior literature if any. At the end of the reaction, the reaction mixture is ready for gel loading.

(v) Experimental Analysis of DNA-Protein Interaction

The sample (reaction mix) can be directly loaded onto the gel. Caution must be adopted in gel loading as the reaction mixture is colorless (and has no loading dye in it), unlike the conventional gel loading where loading dye will be visible. However, the loading of the hot sample can be observed due to the difference in density/viscosity of the glycerol present in the binding buffer. Since there is no loading dye to indicate the gel running (or the front running dye), it is essential to dedicate or use one of the wells of the gel for loading just the dye solution. In that gel lane (well), load 5 µl of the dye/glycerol stock solution (1X), to enable the visualization of gel run.

Electrophoresis may be performed at a low current, 10 mA, and run until the bromophenol blue dye (in the single lane/well as mentioned above) reaches the bottom of the gel or at least 3/4th of the gel. After the completion of electrophoresis, carefully discard the buffers according to the institutional radiation safety guidelines, and remove the gel to fix the protein and its complexes on the gel using the fixative solution. Incubate the gel in the fixative solution for 15 min at room temperature. Next, remove the gel, place it on 3MM Whatman filter paper to enable drying on a gel dryer. Dry the gel at 70 or 80 °C for an hour, before exposing it to the X-ray film for autoradiography. Follow the instructions of the supplier for exposure and development of the X-ray film. Once developed, the X-ray will demonstrate a strong signal at the bottom region (front running dye) in all the wells (lanes) indicative of the unbound, free "hot probe" (free probe), whereas DNA-bound with a protein will show a strong radioactive signal at a higher molecular size depending upon the nature of the protein(s). Shown in Fig. 2.4 is a representative image of EMSA from one of our publications.

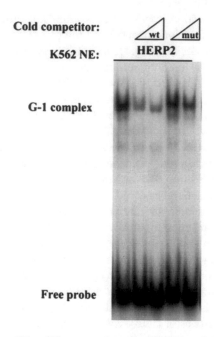

Fig. 2.4 Electrophoretic mobility shift assay to determine the effect of the protein, HERP2 on the interaction between a DNA fragment (nucleotide sequence) that has affinity to bind with the protein, GATA-1 (G-1). In brief, nuclear extract (NE) from the cell line, K562 expressing a recombinant protein, HERP2 was used as the sample to determine the effect of HERP2 on GATA-1 and DNA interaction. The unbound, "hot probe" is identified at the bottom of the gel due to its low molecular mass. The G-1 complex is the DNA fragment and G-1 interaction, and it appears at a higher level (upper region of the gel) indicative of retarded mobility of the G-1 complex due to high molecular mass. The signal of the G-1- complex decreases with increasing concentration of the "cold competitor" (i.e., the DNA fragment without radiolabel, indicated as wild type, Wt). However, when the "cold competitor" (i.e., DNA fragment) was mutated at the G-1 binding site, it did not interfere with the G-1 complex binding (indicated as mut). In conclusion, the presence of HERP2 did not affect the binding of G-1 protein with its corresponding nucleotide sequence (DNA fragment). (Reproduced with permission from Mol Cell Biol (Elagib et al., 2004), American Society for Microbiology)

2.4.6 Study Questions

How is the DNA-binding with a known protein and an unknown protein verified?

2.5 Protein Synthesis

2.5.1 Background

The intricate balance between synthesis and stability determines the functional availability of a protein. In general, cellular proteins have a shelf-life (or life-cycle) after which they undergo programmed breakdown (protein hydrolysis) resulting in the pool of amino acids. The individual amino acids thus released are recycled. Such a regulated degradation of proteins is referred as "protein turnover." The rate of protein turnover, i.e., from protein synthesis to protein hydrolysis, may vary depending upon the specific function of protein species as well as their cellular requirements under normal or pathophysiological conditions. Understanding the rate of synthesis and regulated degradation of a cellular protein in a given condition is pivotal to determine its role in the health and disease conditions of a particular cell.

Current approaches to determine protein turnover involve either direct or indirect perturbation of cellular metabolism. For example, treatment of cells with an inhibitor of global protein synthesis (e.g., antibacterial agents, toxins) will arrest the synthesis of new proteins and will enable the assessment of the stability or degradation of a specific cellular protein already in the cell (before the inhibition of global protein synthesis). Alternatively, a labeling approach may be employed to tag or label the proteins that are synthesized, followed by the isolation or separation of the protein of interest to quantify the amount of labeled protein synthesized and stable (accumulated) in a given cell population. Thus, labeling peptides/polypeptides during protein synthesis, followed by the isolation of a specific protein/peptide of interest, enables quantitation of the rate of synthesis as well as the abundance of the protein in a time-dependent fashion. This approach is valuable in the determination of other aspects of protein chemistry such as protein folding, post-translational modifications, subunit composition, intracellular or subcellular transport, and degradation.

Peptide or protein labeling during synthesis is achieved by the supplementation of the radiolabeled form of an amino acid, particularly the one that is commonly abundant in the majority, if not, all proteins. For example, methionine is a critical amino acid prevalent in all proteins and also serves as the "start codon" for protein synthesis. Hence, ^{35}S-methionine is a widely used radioactive amino acid in protein labeling. There are other labeling methods such as fluorescent probes or fluorescent labels which are also indicated as useful as radiolabeling. Each method of labeling, viz., radioactive isotopes, or fluorescent reporters, has its advantages and disadvantages. In line with the scope of this chapter, we will limit our focus only to the radioisotopic labels used in protein analysis.

The pulse-chase analysis is a well-established and highly adaptable tool for studying the life cycle of endogenous proteins, including their synthesis, folding, subunit assembly, intracellular transport, post-translational processing, and degradation. This unit describes the performance and analysis of a radiolabel pulse-chase experiment for following the folding and cell surface trafficking of a trimeric murine MHC class I glycoprotein. In particular, the unit focuses on the precise timing of pulse-chase experiments to evaluate early/short-time events in protein maturation in cell lines. The advantages and limitations of radiolabel pulse-chase experiments are discussed, and a comprehensive section for troubleshooting is provided. Further, ways to quantitatively represent pulse-chase results are described, and feasible interpretations on protein maturation are suggested. The protocols can be adapted to investigate a variety of proteins that may mature in very different ways.

2.5.2 Method

The widely employed and popular method of choice is the "pulse-chase technique." As the name indicates, the method involves two phases "pulse" and "chase"; during the pulse-phase cells will be exposed to a labeled amino acid so that the protein(s) of interest synthesized will incorporate the labeled amino acid, whereas, in the subsequent chase-phase, the same cells will be exposed to the unlabeled form of the corresponding amino acid to replace or chase the labeled amino acid. Since protein synthesis is a dynamic process, and cells continually synthesize and hydrolyze specific proteins, pulse-chase is a favored method to determine the life cycle of a protein under particular cellular conditions. As mentioned above, ^{35}S-methionine is a precursor of choice for multiple reasons including the radioactive decay properties, the half-life of ^{35}S, etc. In cell culture, physiologically, the equilibrium between intracellular (cytoplasmic) methionine and the supplemented (extracellular) methionine is achieved in less than minutes thus avoiding or minimizing any undesirable delay which in turn will prevent any discrepancy in the estimation of newly synthesized (^{35}S-methionine incorporated) proteins and their programmed hydrolysis.

2.5.3 Principle

The pulse-chase technique is based on the principle that radioactive amino acids can be used to label newly or freshly synthesized polypeptides/proteins. Such radioactive labeling may enable the characterization of the kinetics and fate of the protein of interest.

The principle of pulse-chase is applied to understand the global protein turnover as well as specific protein turnover depending upon the objective of the study. For instance, the effect of a potential therapeutic that targets global translation of cellular mRNA may be verified by investigating the total protein synthesis and hydrolysis. On the other hand, if a specific protein's fate is under investigation, then the pulse-chase involves an additional step of immunoprecipitation or isolation of the specific protein following the pulse-chase. Thus, pulse-chase is a powerful tool to investigate total as well as specific protein in a cell under different conditions (e.g., potential inhibitor, therapeutic, etc.). Note, in the next section, the reagents/equipment mentioned include the list of materials used in the particular study as outlined in the subsequent illustration/example.

2.5.4 Reagents/Equipment

- Cell lines: Human liver cancer cell lines, Hep3B and SKHep1, obtained from the ATCC (Manassas, VA, USA).
- 3-bromopyruvate (3-BrPA) obtained from Sigma Chemical Co. (St Louis, MO, USA).
- All other reagents including NuPAGE Bis-Tris gels, electrophoretic and blot transfer-associated reagents were obtained from Invitrogen (Carlsbad, CA, USA).
- BCA protein assay kit for protein estimation obtained from Pierce Co., (Rockford, IL, USA).
- A refrigerated tabletop centrifuge for protein sample preparation.
- A 37 °C water bath.
- Micropipettes, cell culture reagents, culture wares, sterile mammalian tissue culture hood, incubators, etc.
- The radioactive isotope [^{35}S] labeled amino acid, methionine (^{35}S-methionine) procured from Perkin Elmer, or another authorized commercial supplier.

2.5.5 Stepwise Protocol

(i) The protocol described here uses epithelial cells as prototype, i.e., cells that grow by attachment to the culture vessels also known as the adherent type of cells. For other types of cells such as suspension cells (e.g., blood cells), the method can be adopted with minor modification of initial steps. Figure 2.5 outlines the major steps involved in the pulse experiment described here.

(ii) Cell culture

Hep3B and SK-Hep1 cells were cultured as described by the supplier (ATCC), using a complete growth medium. As in typical in vitro experiments, cells were synchronized a day before the experimentation to maintain a homogeneous population of cells with a similar growth phase (note, cells have growth phase referred to as cell cycle: G1, S, G2, and M-phase). Cells cultured in complete growth medium but with low or reduced serum (1–2% FBS, instead of 10%) may achieve synchronization in general. Drug treatment: As the objective of our experiment is to investigate the effect of the alkylating agent, 3-BrPA on protein synthesis, it is imperative to use a non-toxic, non-lethal dose/condition of the test agent (3-BrPA). At the concentrations (100 μM, 200 μM) used here for the indicated period (60 min) of incubation the cells were viable.

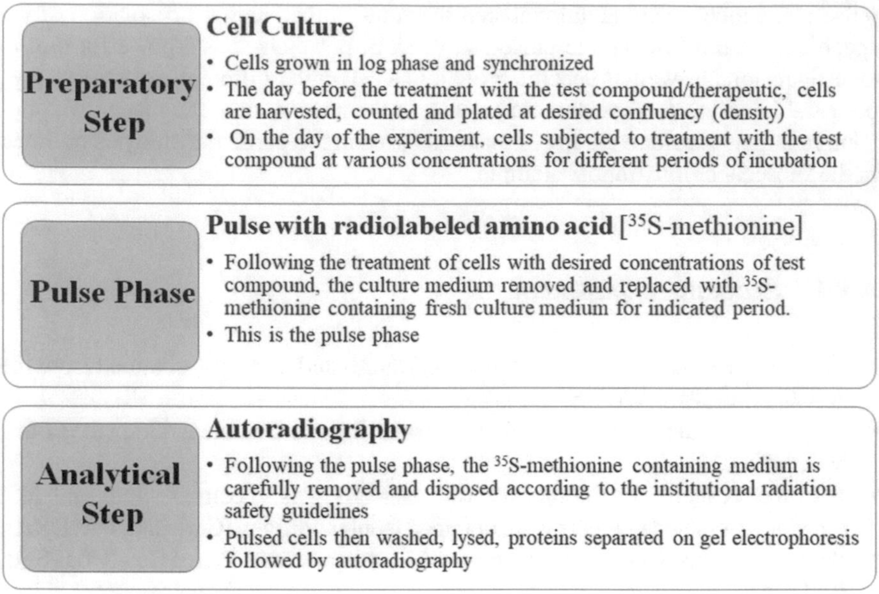

Cell Culture

Preparatory Step
- Cells grown in log phase and synchronized
- The day before the treatment with the test compound/therapeutic, cells are harvested, counted and plated at desired confluency (density)
- On the day of the experiment, cells subjected to treatment with the test compound at various concentrations for different periods of incubation

Pulse with radiolabeled amino acid [^{35}S-methionine]

Pulse Phase
- Following the treatment of cells with desired concentrations of test compound, the culture medium removed and replaced with ^{35}S-methionine containing fresh culture medium for indicated period.
- This is the pulse phase

Autoradiography

Analytical Step
- Following the pulse phase, the ^{35}S-methionine containing medium is carefully removed and disposed according to the institutional radiation safety guidelines
- Pulsed cells then washed, lysed, proteins separated on gel electrophoresis followed by autoradiography

Fig. 2.5 Major steps of a pulse protocol

(iii) Autoradiography and Immunoblotting

The analysis of the incorporation of ^{35}S-methionine into the pool of cellular proteins is performed by autoradiography as described below. Pulse labeling: the day before the experiment, Hep3B and SK-Hep1 cells growing in the log phase were harvested, counted, and plated to achieve 60% confluency. On the day of the pulse experiment, culture medium was removed and replaced with Opti-MEM medium 6 h before the 3-BrPA treatment. After 6 h, the culture cells were treated with either 100 or 200 μM 3-BrPA for 60 min. A batch of untreated cells (without 3-BrPA exposure) served as the control. After 1 h, the culture medium containing 3-BrPA (and also the control) was removed and replaced with [^{35}S] methionine (100 μCi/ml) containing complete growth medium, and the culture was continued for the indicated period (30, 60, and 120 min). For each group, replicates were maintained to validate the findings. At the end of the pulse period, the ^{35}S-methionine containing medium was removed, and the radioactive waste was carefully disposed according to the institutional radiation safety guidelines. The cells were then washed with ice-cold phosphate-buffered saline (PBS), harvested, and lysed in RIPA buffer. After protein estimation, a known quantity of each protein sample was subjected to gel electrophoresis, on a 4–12% NuPAGE Bis-Tris denaturing gel. Duplicate gels were run to use one for protein-staining with colloidal Coomassie brilliant blue (CBB) and the other for autoradiography. For autoradiography, the gels were dried before exposure to the X-ray film (as described elsewhere). The images of the CBB stained gel, as well as the autoradiographic X-ray were scanned (Fig. 2.6).

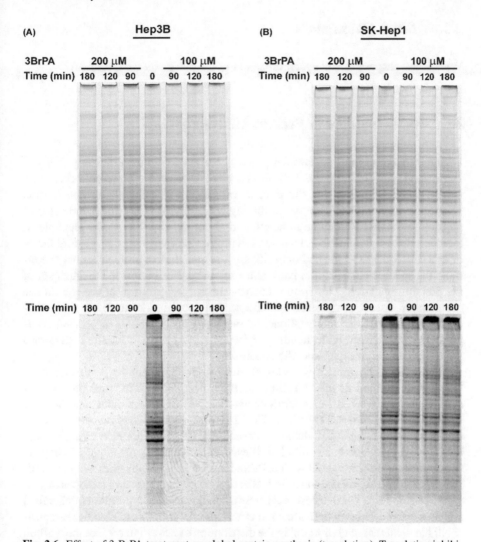

Fig. 2.6 Effect of 3-BrPA treatment on global protein synthesis (translation). Translation inhibition shown by pulse experiment with [^{35}S] methionine. Coomassie-stained gels (top panel) of Hep3B (**a**) and SK-Hep1 (**b**) cells showing equal protein loading. The bottom panel shows the autoradiograms of the corresponding gels. Noteworthy, the newly synthesized proteins will have ^{35}S-methionine incorporation, hence stronger signal on the autoradiogram. However, under the disruption of protein synthesis, ^{35}S-incorporation will diminish hence weak signal on autoradiogram. Intriguingly, Hep3B cells show higher sensitivity to 3-BrPA mediated protein synthesis inhibition as the autoradiogram signal significantly decreased even at 100 μM concentration (bottom panel), compared to the corresponding dose treatment of SK-Hep1 cells. Note, the time indicated in the figure includes the original 3-BrPA exposure period (60 min) followed by continued culture post-exposure for additional 30, 60, and 90 min resulting in the 90(60 + 30), 120 (60 + 60) and 180 (60 + 120) min. (Reproduced with permission from Anticancer Res., (Ganapathy-Kanniappan et al., 2010), IIAR publishers)

2.5.6   Study Questions

What is the purpose of cell synchronization in the pulse-chase experiment?

2.6   Stable Isotopes in Protein Analysis (SILAC)

With the advent of mass spectrometry (MS) combined with the efficiency of liquid chromatography (LC), the LC-MS plays a pivotal role in the proteomic character-ization of biological samples. Quantitative proteomics, i.e., the relative quantitation of specific protein(s) or peptides in biological samples, is a critical indicator of physiology-related alterations in health and abnormal conditions. For example, if there is a disease that involves changes in a protein pathway either as a causal factor or as a consequence of the pathophysiology, the quantitative proteomics may deter-mine the specific changes in a particular protein species and/or the mechanism of the protein regulation. This in turn enables the characterization of the role of the protein(s) in disease progression and/or potential opportunities to develop any ther-apeutic and molecular interventional strategies. Thus, quantitative proteomics is pivotal to advance our understanding of the functional and biochemical alterations related to cellular or organ-specific conditions.

Proteomic estimation of peptides or specific proteins can be performed by a "label-free" method or the "labeling" method. The "label-free" method involves comparative analysis of MS spectra of two samples that are neither modified nor labeled with any reporter chemical. The "labeling" method may involve either (i) the "isotopic labeling" (i.e., stable isotopes) or (ii) the incorporation of a chemical tag. The isotopic labeling in turn is achieved either through metabolic labeling by substitution of an amino acid with a stable isotope (e.g., $^{13}C_6$ lysine) as in cell cul-ture experiments or via chemical derivatization involving the use of isotopes in pro-tein digestion (e.g., ^{18}O in trypsin digestion). Whereas the incorporation of chemical tags is achieved through derivatization of samples/peptides with chemicals/report-ers of identical mass but different in the distribution of heavy (stable isotope) carbon and nitrogen within their structure.

The efficiency of MS depends on the difference in the molecular mass of pep-tides or molecules subjected to analysis. To achieve accurate determination of net change in the abundance or molecular weight of peptides, the MS methodology employs a chemical derivatization approach. In this approach, specific amino acids of peptides or proteins of the samples to be analyzed are modified by the introduc-tion or addition of "stable isotopes" (i.e., the nonradioactive isotopes). Since the stable isotopes (e.g., ^{13}C) that differ from their naturally occurring isotope (e.g., ^{12}C) have higher molecular mass, they are also known as "heavy" isotopes. This distinc-tive increase in molecular mass renders them detectable by MS based on the mass and charge (m/z) ratio. Thus, modification of protein samples includes the addition or introduction of the stable isotope that contains a "heavy" atom, for example, ^{13}C

that differs from the ^{12}C of cellular proteins or peptides. The introduction of such heavy isotopes or tags is achieved through chemical derivatization. Thus, the stable isotopes play a pivotal role in the proteomic analysis by MS. Alternatively, the heavy isotopes may also be introduced into the peptides by the digestion of the protein samples with trypsin in the presence of ^{18}O-labeled water ("heavy" isotope of oxygen). There are two widely used techniques in proteomics in the relative and absolute quantitation of peptides/proteins in biological samples. They are Stable Isotope Labeling by Amino acids in Cell culture (SILAC) and the Isobaric Tags for Relative and Absolute Quantitation (iTRAQ). Here, we will discuss and learn about the relevance and significance of each of these techniques.

2.6.1 Stable Isotope Labeling with Amino Acids in Cell Culture (SILAC)

2.6.2 Background

As the name indicates, the SILAC is relevant for protein analysis of samples in cell culture experiments. SILAC utilizes stable isotopes (nonradioactive) to label specific atoms (e.g., carbon (^{13}C), nitrogen (^{15}N)) of amino acids (e.g., ^{13}C-lysine, ^{13}C, ^{15}N-arginine). As mentioned earlier, the respective amino acids labeled with the stable isotope exhibit an increase in the molecular mass (due to the heavy isotopes ^{13}C or ^{15}N); such amino acids are referred as "heavy" amino acids. On the other hand, proteins of cells cultured in unlabeled amino acids will have low molecular mass compared to the "heavy" (isotope-labeled) amino acids, hence referred as "light" amino acids. In a tryptic-digestion (i.e., enzymatic digestion with trypsin), the same amino acid derived from "heavy" peptide can be distinguished from its "light" counterpart by the difference (increase) in the molecular mass. For example, a ^{13}C-lysine will vary from its "light" counterpart (^{12}C). In other words, SILAC is a metabolic labeling method that involves labeling of cellular proteins with "heavy" amino acids.

Unlike the methods of chemical derivatization which are relevant only in the analysis of existing or available samples acquired from the biological source, SILAC is useful in the characterization of dynamic changes in proteins at various cellular conditions in vivo. In other words, while the chemical derivatization methods are relevant for samples that are at the endpoint of physiology/or pathophysiology, the SILAC technique requires labeling of amino acids in live, metabolically active cells, enabling understanding of protein profile alterations during the progression of a condition or cellular event. Hence the SILAC is aptly known as the metabolic labeling technique.

2.6.3   Method

The application of SILAC is to identify or determine changes in the quantity of a protein or proteins in a given cellular condition. The experimental condition may include a wide range of options from cell growth and development to the pathogenic transformation of normal cells. In other words, SILAC enables protein analysis by comparison of a healthy or control (natural) sample with the abnormal/experimentally perturbed condition. Thus, control cells cultured in a regular growth medium (often referred as "light" amino acid) will be compared with the experimental cells cultured in a growth medium containing one or two "heavy" amino acids. A diagrammatic representation of the overall method of SILAC is shown in Fig. 2.7a. At the end of the experiment, samples prepared and subjected to MS analysis to determine any change in the abundance of a specific peptide/protein (Fig. 2.7b).

Prior to the experimental design, it is imperative to comprehend several key principles and cautionary notes for efficient sample analysis.

The choice of amino acid used in the preparation of "heavy" culture medium depends on factors such as "essential" or "non-essential" amino acids. The essential amino acids are required to be supplemented via the growth medium and, are not obtained through diet or other processes (e.g., methionine, lysine). On the other hand, "non-essential" amino acids (e.g., tyrosine, arginine) are obtained via nutrition/diet at the organismal level. In cell culture, it is known that arginine is indispensable for cell growth and survival. Accordingly, the commonly used amino acids for "heavy" labeling are lysine and arginine, ^{13}C-lysine, and ^{13}C, ^{15}N-arginine.

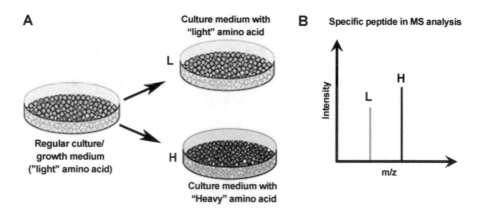

Fig. 2.7 Diagrammatic representation of key steps of SILAC experiment. *Panel A*, Cells are cultured in a regular growth medium with unlabeled ("light") amino acid. On the day of the experiment, cells growing in the log phase are divided or split into two for subsequent culture in the presence (H) of isotope-labeled ("heavy") amino acid or unlabeled ("light", L) amino acid. *Panel B*, schematic showing MS-identification of the difference in the intensity/abundance of a particular peptide based on the difference in mass attributed to the isotope-labeled amino acid

Noteworthy, eventually the sample preparation for MS analysis involves trypsin digestion to generate peptide fragments. Since trypsin is a very specific proteolytic enzyme and cleaves the carboxy-terminal (C-terminal) of lysine or arginine amino acids, these amino acids are widely used in SILAC.

Next, the rate of incorporation of the labeled amino acid into the global protein profile of cells is directly proportional to the rate of cellular requirements for protein synthesis. Thus, to achieve near homogeneous labeling of all the proteins of cells with "heavy" amino acid, the cells must be allowed to replicate, grow, and repeat the process for a few cycles. In other words, culturing cells in a "heavy" medium for at least a few generations (e.g., 4–5 replications) may be essential to achieve complete labeling of all the proteins of a cell population. It is essential to determine the rate of replication of cells, and the duration of culture in the "heavy" medium must be empirically determined before the actual experimental analysis of samples.

Finally, and most importantly, the cardinal principle of SILAC experiment is the supplement of the specific amino acid with a "heavy" isotope as the only source of the particular residue in the cell culture/medium. In other words, to obtain flawless, reliable data and interpretation, the "heavy" amino acid must be the only source of supplement for the referred amino acid. For example, if ^{13}C-lysine is used as the "heavy" amino acid in SILAC, then the cell culture condition (including growth medium) must be devoid of the "light" lysine, i.e., the natural isotope (^{12}C) of lysine. This is critical; otherwise, the presence of "light" and "heavy" in the same culture medium will result in the generation of mixed type of labeling, with "heavy" proteins and "light" proteins leading to erroneous quantification. One of the major sources of contamination (i.e., source of "light" amino acid) is the serum-supplemented in the cell culture medium. This can be addressed by several approaches such as dialysis of serum (to remove the naturally occurring amino acids) followed by manual addition of the individual amino acids (including the "heavy"). However, dialysis may result in the elimination of some essential growth factors (e.g., insulin) as well, which may affect cell growth or development. Arguably, if the same dialyzed serum is used for the control cells (with the light amino acid), then the effects of dialysis-related changes must be common in both control and experimental (with the "heavy" isotope). However, this is not a biologically sound approach, as the changes in cellular characteristics may undermine the overall objective of the study. Hence, it is recommended that following the serum dialysis, growth factors be manually added to the culture medium, to reconstitute a complete growth medium irrespective of the "heavy" or "light" amino acid. SILAC experiment necessitates three broadly classified critical steps to achieve biologically valid results. The steps include a preparatory step, confirmatory step, and the final experimental analysis (Fig. 2.8).

Several commercial kits are available for the SILAC labeling of cells cultured in a specific growth medium.

Preparatory Step	Confirmatory Step	Experimental analysis
• Preparation of cell culture medium with labeled ("heavy") amino acid and "light" (natural isotope) amino acid. • Optimization of cell culture with growth factors or any necessary supplements to be added to the "heavy" and "light" growth medium. • Confirmation of cell growth and replication	• Verification of the efficiency of isotope incorporation of the labeled amino acid in global protein profile. • Determination of the duration of cell culture required to achieve >95% incorporation of the isotope of the labeled amino acid in the protein(s).	• Experimental design to investigate the desired objective/hypothesis. • Typical study ranges from the analysis of cellular response to cytotoxic agents/ therapeutics to understanding cell differentiation and apoptosis and so on. • Prepare the samples and analyze by MS to identify changes in specific protein(s).

Fig. 2.8 Major steps of a SILAC protocol

2.6.4 Principle

SILAC relies on the fundamental principle that protein synthesis is an integral part of viable, metabolically active cells. Thus, culturing cells in the presence of labeled-amino acids will lead to the synthesis of cellular proteins with the specific label. In SILAC, the label used is an amino acid with a stable isotope (e.g., ^{13}C-lysine) which differs only in the molecular mass of the particular atom that does not affect the structural, biochemical, or functional properties of the proteins synthesized. Nevertheless, the introduction of the stable isotope will enable us to differentiate the proteins in MS analysis due to the increase in molecular mass associated with the "stable isotope" used in the experiment. Since, any change in cellular physiology will involve alterations in the abundance and/or nature of proteins synthesized, by comparison between two cell populations (e.g., control and experimental) the difference in a particular protein of interest or unknown proteins may be distinguished by the MS analysis.

2.6.5 Reagents/Equipment

- Cell line (e.g., Hep3B).
- Cell culture medium (base) (e.g., DMEM, RPMI) as required for the cell line.
- FBS, fetal bovine serum, dialyzed FBS.
- Stable isotope-labeled amino acid, "heavy" amino acid (e.g., ^{13}C$_6$ L-lysine).
- Phosphate buffered saline (PBS) ~ pH 7.2.
- Cell lysis buffer (e.g., RIPA lysis buffer).

- Protein estimation kit (e.g., commercially available kits such as BCA assay kit).
- In-gel tryptic digestion kit (e.g., commercially available trypsin for digestion).
- Gel electrophoretic apparatus, reagents including SDS-PAGE gels (e.g., Bis-Tris gel), buffers, etc.
- Gel staining reagent (e.g., Coomassie Brilliant Blue, CBB).
- A refrigerated tabletop centrifuge for protein sample preparation.
- A 37 °C water bath.
- Micropipettes, cell culture reagents, culture wares, sterile mammalian tissue culture hood, incubators, etc.

2.6.6 Stepwise Protocol

(i) Preparatory Step

Preparation of cell culture medium with labeled ("heavy") amino acid.

Following the instructions of the cell line provider, and the media supplier, prepare the cell culture medium with dialyzed-FBS, to achieve desired volume and composition (e.g., 10% FBS) of the media components. Next, add the "heavy" amino acid, to obtain the recommended concentration in the culture medium. For example, if the amino acid L-lysine (natural isotopic) is recommended or present at 1 mM in the regular, complete growth medium, then the "heavy" amino acid ($^{13}C_6$-lysine) must be added to achieve the final 1 mM concentration.

Next, to compensate for the loss of any specific/essential growth factors in the process of FBS dialysis, add such factors manually to make up the complete growth medium with the "heavy" amino acid. Depending upon the cell line, other supplements including antibiotics may be added as required. As with the majority of the cell culture media, protect the culture medium from light and store at 4 °C until use.

Finally, test the growth efficiency and cell culture characteristics of the cell line in the presence of the "heavy" medium and compare with the cells that are cultured in a regular, complete growth medium ("light"). Note, "light" refers to a naturally occurring isotope which is the regular amino acid and not labeled with any heavy-isotope. This will enable us to confirm if the cell growth and development are normal and unaffected by the "heavy" medium.

(ii) Confirmatory Step

In this step, the incorporation of the isotopic labeled-amino acid into the cellular proteins will be verified to assess the efficiency as well as to determine the duration of culture required to achieve the incorporation. It is desired that at least >95% of the protein(s) exhibit the incorporation of labeled-amino acid.

Start with culturing cells in the log phase and determine the doubling time (i.e., rate of replication). Once the cell density (also known as confluency) reaches between 60% and 80%, split the cells into two and culture each one with "heavy" or "light" culture medium representing "isotope-labeled" and "natural amino acid," respectively.

Culture the cells in their respective "heavy" or "light" medium for 4–5 generations (doublings). Splitting (subculture) may be performed only when the cells are growing in the log phase (i.e.) at least 60–80% confluent (cell density). Neither the over-crowded (super confluent) nor the diluted (poor confluent) cells are recommended as they affect the biological properties of cells.

Next, after culturing the cells in the respective "heavy" or "light" medium, for 4–5 generations, harvest a portion of cells. For example, if the culture dish (vessel) has 5×10^6 cells, harvest 1×10^6 cells, for experimental evaluation of the isotope incorporation efficiency. The remaining cells may be maintained for subsequent culture. Or if desired, the remaining population of cells from the respective culture may be "cryopreserved" and stored for future use.

Evaluation of isotopic amino acid incorporation is performed by cell lysis using appropriate buffers (e.g., RIPA); estimate the protein concentration. Next, protein samples are prepared and resolved by SDS-PAGE electrophoresis. Each of the samples ("heavy" or "light") will be subjected to SDS-PAGE electrophoresis. At the end of electrophoresis, stain the gel with protein staining reagent (e.g., CBB), and excise (cut) the same protein band from the respective samples ("heavy" or "light") gels (or lanes). Digest the excised protein bands with MS-compatible Tryptic digestion and subject to MS analysis. The amount of incorporation of amino acid labeled with the isotope can be determined by the shift in the spectrum and its peak intensity (e.g., Fig. 2.7b). The results will indicate whether the duration of culture in the "heavy" medium is sufficient or requires a longer culture to increase the incorporation efficiency. If incorporation of the isotope is >95%, the experimental analysis (study) may be performed using the cell population (in culture or cryopreservation, as indicated above (Confirmatory step)). If the isotope incorporation is low, then the cells need to be cultured for additional time to achieve >95% incorporation. Nevertheless, the efficiency of incorporation needs to be done again until it reaches >95% incorporation of the heavy isotope.

(iii) Experimental Analysis

As mentioned earlier, the SILAC experiment may be employed to investigate changes in a protein(s) during cellular response to changes in the internal or external milieu, for example, changes in a protein during (a) cell differentiation, (b) disease progression, (c) specific gene-silencing, (d) exposure to cytotoxic agents or therapeutics, and so on.

In the experimental investigation, the desired objective or hypothesis may be tested before the harvest and cell lysis. Following the cell lysis, the analysis of proteins is performed as described above using gel electrophoresis, in-gel-digestion of the specific protein, and MS-analysis. Noteworthy, the protein analysis can be streamlined to specific subcellular compartments (e.g., nuclear proteins, mitochondrial proteins, membrane proteins) by optimizing the method of cell lysis, and sample preparation. Several, commercially available kits provide specific reagents and instructions on the isolation of proteins from specific subcellular compartments which will avoid cross-contamination from the rest of the cellular protein pool.

2.6.7   Study Questions

(i) In the SILAC experiment, why is it necessary to dialyze the FBS to be used in the cell culture medium?
(ii) From the biological sample perspective, what is the limitation of the SILAC approach in protein analysis?

2.7   Stable Isotopes in Protein Analysis (iTRAQ)

2.7.1   Isobaric Tags for Relative and Absolute Quantitation (iTRAQ)

2.7.2   Background

The iTRAQ is a widely used proteomic method in the clinical and preclinical tissue analysis encompassing various aspects from biomarker studies to molecular mechanisms and therapeutic targets. Quantification of peptides using isobaric tags is a commonly adopted method in proteomics. One of the distinctive advantages of iTRAQ, besides the ability to achieve both relative and absolute quantification of peptides, is the ability to multiplex sample analysis. While other methods of chemical-derivatization such as ^{18}O (during digestion) or ^{13}C-acrylamide also achieve the incorporation of tags to peptides, they are applicable only in binary comparisons (two samples). Whereas, with iTRAQ using a 4-channel or 8-channel approach, up to eight samples can be investigated simultaneously, thus reducing or eliminating handling variability. Another major advantage of iTRAQ is its utility in labeling the samples from any source. This is very relevant and useful in the clinical investigation of patient samples including body fluids and tissues of various pathophysiology. Unlike the SILAC, a metabolic labeling method that requires cell lines treated with "heavy" vs "light" isotopes, the iTRAQ technique can be applied to already acquired/harvested tissues or samples. With the advancement in multiplex techniques, and the relevance in the analysis of tissues and fluids from patients and preclinical models of human disease the iTRAQ application has gained significant attention.

2.7.3   Method

As mentioned earlier, the efficiency of iTRAQ analysis relies on several critical steps from sample preparation to final LC-MS/MS analysis. The overall method can be broadly classified into three steps as described below.

 (i) Sample Preparation and Protein Digestion

As indicated in the "Background" section the salient feature of iTRAQ method is the utilization of biological samples from any source including blood plasma/serum, tissues or cell lines, and so on. Nevertheless, the preparation of protein samples requires careful application of buffers and other reagents that are devoid of any chemical components that may interfere with downstream application such as labeling. Specifically, as indicated elsewhere, the presence of free amine groups, or chemicals that may donate such amine groups could interfere with iTRAQ labeling. Thus, the samples and the buffers need to be clarified of such chemical interference by careful preparation or dialysis. Following the sample preparation, the proteins in the sample will be subjected to trypsin digestion as outlined in the stepwise protocol. The digested protein, i.e., the peptides of the samples, are then subjected to iTRAQ labeling. At this point, if necessary the samples may be stored at −20 °C upto a month before further processing. However, it is highly recommended to proceed to the next step to achieve labeling, after which the samples may be stored for future use.

 (ii) iTRAQ Labeling

Following the preparation of peptide samples by digestion, the next step is to label the peptides with respective iTRAQ agents. As outlined earlier, labeling can be multiplexed to achieve simultaneous labeling of 4 or 8 samples with different reporter tags. The labeling instructions are provided by the supplier of the iTRAQ reagents. In general, the step involves careful removal of an aliquot of respective tags from the vials followed by its addition to the tubes containing peptide samples. The labeling reaction begins immediately upon the addition of the tag; hence further incubation of the reaction mixture (i.e., tags plus samples) for 1–2 h is sufficient to complete the labeling of the entire mixture. Each sample will be tagged with one iTRAQ reagent; thus up to 4 or 8 samples can be labeled and analyzed concurrently. Following the labeling reaction, the samples may be used for downstream analysis or stored at −20 °C for 2–3 months.

(iii) Chromatographic Fractionation and Mass Spectrometric Analysis

The final step involves the actual separation of peptides by high-performance liquid chromatography (HPLC or LC) followed by tandem MS (MS/MS) analysis.

2.7.4   Principle

The principle underlying the iTRAQ is that labeling of peptides with isobaric tags at the N-terminus and the ε side chain of lysine residues differentiate the fragments of peptides based on the distribution of ^{13}C and ^{15}N (heavy isotopes) of the tag. Precisely, in the iTRAQ tags, N-hydroxysuccinimide acts as the reactive group that labels the target peptides through the free amine group at the N-terminus or the side chains of lysine residues. Thus, efficient peptide labeling by iTRAQ tags relies on the free amine group. Noteworthy, due to the critical requirement of free amine groups for labeling, the buffers used for sample processing especially in the labeling reactions mustn't contain free amines or amine groups (e.g., ammonium bicarbonate, Tris). Otherwise, the amine groups present in the sample/buffer may compete and impede the iTRAQ labeling. The elegance and robustness of iTRAQ is also due to the principle of applying tandem MS, i.e., MS/MS, as contrary to the single MS used in other proteomic methods.

Chemically, each tag used in the iTRAQ consists of three components: a low-mass reporter group, a balance group, and an amine-reactive group. They have isobaric masses in the MS mode. However, upon fragmentation, the tags release the low-reporter ions allowing the quantification at MS/MS level. Typically, in the 4-channel or 4-plex, the low-mass reporter ions are 114.1, 115.1, 116.1, and 117.1 (m/z). Whereas in the 8-plex or 8-channel method, additional four low-mass reporter ions such as 113.1, 118.1, 119.1, and 121.1 (m/z) are added to the 4-plex ions. Thus, the application of MS/MS enables absolute and relative quantification of peptides by iTRAQ.

2.7.5 Reagents/Equipment

- Phosphate-buffered saline (PBS; 1X).
- Radioimmunoprecipitation assay (RIPA) buffer, [150 mM NaCl, 1.0% IGEPAL, 0.5% sodium deoxycholate, 0.1% SDS, and 50 mM Tris, pH 8.0].
- BSA, bovine serum albumin (Sigma Aldrich, USA).
- iTRAQ reagents from commercial suppliers (e.g., Sciex Inc., MA, USA).
- Triethylammonium bicarbonate (TEAB) buffer (1 M, pH 8.2) (Sigma-Aldrich, USA).
- SDS, Sodium dodecyl sulfate (2%) from iTRAQ reagent supplier.
- Tris-(2-carboxyethyl) phosphine (TCEP) (Sigma, USA). A stock of 50 mM is prepared in double distilled water and stored at −20 °C.
- Methyl methanethiosulfate (MMTS) (Thermo Scientific, USA). A stock of 200 mM is prepared in isopropanol and stored at −20 °C.
- Mass Spectrometry grade Trypsin (Promega, USA).
- Protease inhibitor cocktail-EDTA free: Mixture of several protease inhibitors inhibiting serine, cysteine, but not metalloproteases.
- Refrigerated table-top microcentrifuge.
- Protein estimation assay kit (e.g., BCA assay kit), (ThermoFisher Scientific, USA).
- All reagents must be of analytical grade. Basic equipment such as pipettes, and centrifuges as described for the SILAC are relevant in iTRAQ as well.
- The exhaustive list of HPLC reagents, Mass Spectrometry reagents, and the instrument details are beyond the scope of the book (i.e., tracer technique) as those instrument-based analytical techniques are common for tracer and non-tracer studies.

2.7.6   Stepwise Protocol

(i) Sample Preparation

Depending upon the source of the sample, the preparation method and choice of buffer need to be carefully selected. For example, if the sample source is blood plasma or serum, there is no requirement of homogenization or any cell lysis buffer. However, if the biological source is a cell line or tissues, then the cell/tissue lysis may be performed using ice-cold RIPA buffer or PBS buffer, followed by repeated centrifugation to separate hard, insoluble debris and the clear supernatant with proteins. This clarified sample will be the starting material for further processing. To avoid any inadvertent introduction of free-amine groups from the buffers used in sample preparation, it is desirable to use the TEAB buffer (1 M) with 0.1% SDS, as this buffer efficiently lyses majority of cells.

Next, if organelle (e.g., mitochondria, nucleus) specific protein analysis is performed, then careful separation and isolation of the organelle is critical to prevent cross-contamination of proteins from other subcellular compartments. Care must be taken to avoid proteolytic degradation of sample proteins by cellular/tissue proteases. Hence, the use of protease inhibitor cocktail-EDTA-free is recommended in the lysis buffer (e.g., RIPA, PBS). However, following the sample preparation, these inhibitors need to be dialyzed and removed to avoid any potential interference in subsequent steps (e.g., digestion).

(ii) Protein Estimation

Once the samples are ready for analysis, first and foremost is the quantification of proteins to determine protein concentration of samples. Conventional protein estimation methods (e.g., BCA protein assay kit) would suffice. However, the 2D-quant kit from any supplier (e.g., Millipore Sigma, USA) is recommended due to the nature of the protocols which eliminates potential interference of chemical constituents of sample buffers. The absolute quantity/concentration of the protein may be determined using appropriate reference-standard (e.g., BSA). Following the estimation of protein, aliquot a known quantity (e.g., 100 μg) of each protein sample into fresh Eppendorf tubes, and equalize the volume of each of the samples by the addition of TEAB buffer. In other words, adjust the volume of each sample to achieve an equal amount of protein in the same volume. Ideally, the final volume of protein sample be between 20 and 40 μl for each sample. If the volume is high, it can be reduced by a vacuum dryer (e.g., SpeedVac concentrator). At this point the samples may be processed further (e.g., digestion) or stored at −80 °C. If storage is preferred, then dry the samples in a vacuum dryer (e.g., SpeedVac concentrator) at 45 °C, and the dried samples may be stored at −80 °C before digestion.

(iii) Proteolytic Digestion

The samples of known protein quantity in equal volume of the buffer may then be subjected to a series of steps prior to digestion. Note, if using frozen/stored samples, they must be brought to RT (room temperature) from −80 °C, and dissolved in 40 μl of TEAB buffer. To enhance the solubilization, samples may be thoroughly vortexed, and if necessary 2–3 μl of 2% SDS may be added to the TEAB. The pre-digestion process includes reducing the disulfide bonds of cysteine residue followed by the alkylation of the reduced cysteine residues. First, incubate the samples (~40 μl) with 4 μl of TCEP (i.e., 1/10th of the volume of sample), vortex, pulse spin, and incubate for 1 h at 60 °C. Next, add 2 μl of MMTS (i.e., 1/20th of the volume of sample), vortex, pulse spin, and incubate at RT for 10 min. Following the reactions of TCEP and MMTS, the samples are subjected to trypsin digestion. Add 10 μg of trypsin (at a concentration of 1 μg/μl in TEAB buffer), vortex, pulse spin, and incubate at 37 °C overnight (~12–16 h). Note, efficient trypsin digestion requires that the reaction mixture doesn't have any interfering substances at inhibiting concentration. For example, SDS concentration must be maintained ≤0.05% of the total reaction volume.

(iv) iTRAQ Labeling

Following the trypsin digestion, if required, reduce the sample volume by SpeedVac contractor at RT (avoid complete drying or desiccation as it may affect the peptides), and adjust all samples to equal volume (e.g., 30 or 40 μl). At this point, if necessary the samples may be stored at −20 °C for up to 1 month, although it is recommended to proceed with the labeling.

For the labeling reaction, remove the required iTRAQ labels from the kit supplied by the manufacturer. From this point, act quickly to spin down the labeling reagent vial to collect the contents to the bottom of the tube. The thawing should be swift and not more than 2–3 min. Next, add 60 μl of ethanol or isopropanol depending upon the iTRAQ 4-plex or 8-plex reagent kit, respectively. Briefly mix, and pulse spin. Next, transfer the contents of the labeling reagent vial to appropriate sample tubes, and rinse with 10 μl of either ethanol or isopropanol accordingly. Now, mix the contents of the sample tube containing the labeling reagent by vortex mixer, and allow the labeling reaction by incubation at RT for 1 or 2 h depending upon the 4-plex or 8-plex reagent, respectively.

Finally, following the labeling reaction, reduce the sample volume to ~30 to 40 μl by SpeedVac concentrator at RT; care must be taken not to dry the samples. At this point, the samples may be stored at −20 °C for 2–3 months.

(v) Fractionation and Quantification by Liquid Chromatography-MS/MS

Once the iTRAQ labeling is completed, the samples are ready for peptide fractionation by chromatography (e.g., HPLC), followed by ionization and quantification of individual peptides/proteins by tandem MS (i.e., MS/MS).

The isotope labeling (i.e.) the tracer component of the entire method is defined until the section of iTRAQ labeling. Subsequent steps such as chromatographic separation and MS analysis are common methodologies and are not limited to tracer studies, hence not covered in this chapter. The reader is recommended to refer to any source of material relevant to common chromatographic and MS methodologies.

2.7.7 Study Questions

 (i) What is the distinctive feature of iTRAQ that renders its application relevant compared to SILAC?
(ii) In the iTRAQ labeling, why is the presence of free amines a critical concern?

Chapter 3
Tracer Technique in Applied Research

3.1 Introduction

Clinical translation of a potential drug (e.g., anticancer agent) is guided by two cardinal principles, efficacy and safety. Efficacy refers to the likely outcome or benefits, whereas safety corresponds to the level of undesirable side effects or toxicity. Essentially, the efficacy depends upon how the body responds to the drug (pharmacodynamics) while safety relies on how the drug is absorbed, distributed, metabolized, and excreted (ADME) also referred as pharmacokinetics. Failure in the translation of majority of new chemical entities is due to the lack of complete understanding of their pharmacokinetics and drug metabolism [Roberts 2003]. Thus, a good "candidate drug" needs to be assessed for its distribution and availability. Since alkylating agents pose challenges in the detection and estimation as described elsewhere, radiolabeling proves to be a valuable approach.

3.2 Pharmacokinetics

3.2.1 Background

Despite the characterization of the specific molecular mechanism of action at the intracellular level, several antineoplastic alkylating agents react with off-targets, such as serum proteins, during systemic administration. As discussed elsewhere such nonspecific interactions or alkylation form a major impediment in the preclinical evaluation of pharmacokinetics. Hence, a robust, reliable yet widely accepted methodology is required for scientifically sound pharmacokinetic analysis of challenging compounds such as halogenated agents.

© Springer Nature Switzerland AG 2022
S. Ganapathy-Kanniappan, *Isotopic Tracer Techniques in Preclinical Research*,
Techniques in Life Science and Biomedicine for the Non-Expert,
https://doi.org/10.1007/978-3-030-99700-7_3

In the past few years, tremendous advancements have been made in the analytical methods, particularly in the pharmacokinetic analysis of antineoplastic alkylating agents (e.g., halogenated compounds). However, the chemical modification of such halogenated alkylating agents upon interaction with their molecular targets remains a primary challenge in the estimation of the potential drug or compound. This chemical modification, i.e., alkylation, thwarts detection and/or quantification of the parent compound (i.e., the potential anticancer agent). In this chapter, we describe the relevance and application of the tracer technique [which employs radioactive isotopes as labeling molecules] in pharmacokinetic studies. Here we outline a simple protocol that can be adopted to investigate the distribution and metabolism of any potential antineoplastic halogenated compound in vivo. The use of ^{14}C-labeled compounds is described as an example, but other isotopes (e.g., ^{3}H) can also be used. In brief, the radiolabeled compound of interest (for simplicity referred to herein as ^{14}C-R) is administered via intravenous or systemic delivery. The bioavailability of ^{14}C-R is determined by counting ^{14}C in serum and/or plasma using a scintillation (beta) counter. Verification of interaction or alkylation of ^{14}C-R with blood components is performed either by protein precipitation followed by scintillation counting or by gel-autoradiography of serum proteins. The biodistribution of ^{14}C-R is determined by analyzing tissues of various organs either by liquid scintillation counting of tissue homogenates, or gel-autoradiography, or tissue-autoradiography. Finally, the excretion of ^{14}C-R can be assessed by the scintillation counting of urine collected from the bladder. Importantly, the tissue-autoradiography of the brain can demonstrate if the ^{14}C-R can penetrate the blood-brain barrier and has the potential to cause neuronal toxicity. This is very significant as this will enable us to establish dosage profile and route of administration to achieve patient safety. Although radiolabeled compounds are costly, resource-intensive, and necessitate the implementation of radiation safety guidelines, yet the accuracy and reliability of data obtained by tracer technique underscore its significance in the preclinical assessment of emerging anticancer alkylating agents.

3.2.2   Method

The present protocol involves the use of the radiolabeled compound (e.g., ^{14}C-R) and experimental animal model (e.g., rat). Hence, appropriate institutional guidelines that govern these procedures must be followed. Appropriate radioactive decontaminations/containments must be followed strictly according to the Safety Rules and Regulations of respective institutions. Only the personnel authorized and trained to handle radioactive materials should perform all the procedures. Similarly, all animal study protocols must be approved by the Institutional Animal Care and Use Committee (IACUC). Animal experiments may be performed as per institutional guidelines in dedicated animal facilities. All animals must be handled according to the guidelines of NIH and protocols approved by the IACUC.

3.2.3 Principle

A brief description of a representative halogenated anticancer agent, 3-bromopyruvate (3-BrPA), is provided here to enable readers from diverse background to better understand the chemical intricacies. Figure 3.1 shows a schematic diagram of 3-BrPA, a halogenated (bromine) pyruvate.

As shown in Fig. 3.1, in the presence of any electron acceptor (e.g., H+), bromine (being an active leaving group) gets released from the pyruvate backbone and forms a halogen-bond with the electron acceptor (e.g., H-Br). This eventually results in the generation of an unstable electron-deficient (electrophilic) pyruvate backbone. In other words, the release of bromine leads to the generation of the reactive-pyruvate backbone (X) which readily binds with an electron donor, a process known as alkylation (X-pyruvate backbone). Thus, pharmacokinetic analysis of a bromo-compound under preclinical investigation is convoluted. Nevertheless, if the compound is labeled with a radioisotope (e.g., ^{14}C), it will be possible to trace the presence and distribution of the radiolabeled material (despite being alkylated with targets molecules). Noteworthy, it is imperative to label the principal chemical moiety that induces the anticancer effects. In the case of 3-BrPA, it is the alkylation by electrophilic pyruvate backbone that binds with the intracellular target, GAPDH, a glycolytic enzyme. Hence, the carbon atom in pyruvate must be labeled with ^{14}C. In this report, we describe a detailed protocol that can be adopted to investigate the absorption, distribution, metabolism, and excretion (ADME) in vivo of an antineoplastic alkylating agent of similar class [Kunjithapatham et al. 2013].

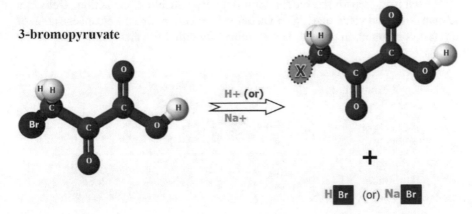

Fig. 3.1 Schematic showing the chemical structure and alkylation property of 3-bromopyruvate (3-BrPA) used in the tracer studies. One of the carbon atoms [C] is labeled with the radioisotope, ^{14}C

3.2.4   Reagents/Equipment

- Radiolabeled compound (e.g., ^{14}C-3-BrPA).
- Radiolabeled compounds can be custom synthesized if the investigator has an authorized, well-equipped laboratory or the radiolabeled compound [^{14}C-R] can be procured from commercial sources. The representative results shown in this protocol used ^{14}C-3-BrPA procured from Perkin Elmer Inc.
- Experimental model (e.g., Sprague Dawley rats).
- The pharmacokinetic study can use any animal model which suits the expertise, interest, and facilities available with the principal investigator. The representative results shown in this report involved the use of male Sprague Dawley rats. Since the focus of the study is to assess the ADME and not the therapeutic efficacy, the animals used here were healthy, normal rats and did not have any tumor.
- Electrophoretic chemicals and reagents as described elsewhere.
- Chemicals/stains specific for histopathology and histochemistry.
- Autoradiography requirements such as X-ray film, and film processor/developer.
- Tissue homogenizer.
- Electrophoretic apparatus and powerpack.
- Microcentrifuge, table-top refrigerated centrifuge.

3.2.5   Stepwise Protocol

The mode of administration and the blood sample collection routes can be optimized as per the investigator's expertise and interest as approved by the corresponding IACUC. The method described herein is what we adopted but in no way meant to emphasize or suggest this as the only approach. A schematic of the workflow or outline is indicated in Fig. 3.2.

Restrain or prepare the rat for drug delivery and blood collection. Deliver a known dose (and volume) of ^{14}C-R via tail vein injection. The dose/concentration of the test compound, in general, is determined by either the therapeutic dose or the

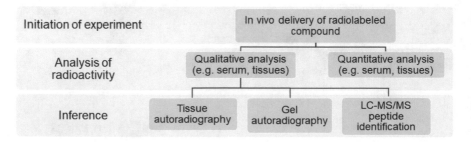

Fig. 3.2 Schematic showing the principal steps of pharmacokinetic analysis of a radiolabeled compound

maximum tolerated dose. The volume of the dose is determined based on the principle that the administered drug volume does not exceed 3.5% (i.e., the median of 2–5%) of the total rat-blood volume.

3.2.5.1 Sample Collection/Preparation

Blood Collection and Fractionation

Collect blood samples from rats at different time intervals according to the protocol approved by IACUC. For the preparation of serum samples, initiate serum formation by induction of coagulation at room temperature for 30 min followed by centrifugation at $1200 \times g$ for 15 min at 4 °C. Separate the clear serum and store at −80 °C until further use. Rinse the pellets with wash buffer [ice cold PBS] and centrifuge to obtain a clean pellet. Remove/decant the wash buffer thoroughly and store the pellets at −80 °C. On the other hand, if blood plasma is preferred over serum, then collect the blood in an anticoagulant tube (to prevent clotting), centrifuge at 1500 to $2000 \times g$ for 20 min in a refrigerated centrifuge at 4° C. Carefully remove and transfer the clear supernatant (i.e.) the plasma to fresh tube and store at −20° C (for short-term storage) or −80° C until further use.

Bladder Puncture-Urine Collection

Urine represents one of the critical indicators of elimination of the test agent (potential anticancer agent) via excretion. Hence, before harvesting vital organs, carefully puncture the urinary bladder using a fine syringe and collect the urine, and store at −20 °C for further analysis. A brief centrifugation step may be performed to clear the urine of any undesirable particulate materials (such as the occasional presence of cellular materials). If a kinetic study is required regarding the frequency and quantum of excretion of the investigational agent (^{14}C-R), then ideally urine must be collected from the animal at desired time intervals in which case specialized or custom-made animals cages that are approved by IACUC may be used.

Tissue Harvest and Fixation

Euthanize rats by the method as approved by IACUC (in general, either carbon dioxide exposure or cervical dislocation are preferred owing to the quick and immediate arrest). Care must be taken to complete the entire tissue harvest procedure as quickly as possible to avoid any decomposition or unwanted degradation. Harvest organs such as heart, lung, liver, kidney, and brain, and excise a portion of each organ into slices (small pieces) of ~1–2 cm^3 and fix immediately in phosphate-buffered 10% formalin (Polysciences Co., Warrington, PA). Fix the tissues in formalin between 3 days to a week before the preparation of tissues for histology.

Tissue Snap Freezing

Prepare small pieces (slices) of tissues from the organs as described above and quickly transfer to cryovials or Eppendorf tubes and plunge them into liquid nitrogen. This will achieve quick and complete freezing of tissue for downstream analysis. In case liquid nitrogen is not available, a dry-ice-ethanol bath can be used which enables us to freeze tissues quickly. Such frozen tissues must be stored at −80 °C until further use.

3.2.5.2 Quantitation of Radioisotope

Scintillation Counting

The advantage of the systemic administration of the radiolabeled compound (^{14}C-R) is that its presence can be evaluated in blood and plasma or serum proteins as well. The presence of ^{14}C-R in total plasma or serum (in microliters) or in respective protein fractions can be assessed by their addition into scintillation cocktail (a fluid that has a solvent, emulsifier, and fluor) followed by the measurement of beta emission in a counter. Add a known volume (100 μl or more) of total plasma or serum sample to 4–5 ml of scintillation cocktail (Cytoscint, catalog no. 882453; MP Biomedicals, CA) and record the readings in a beta counter such as Beckman Coulter LS6500 or similar counter. For the analysis of respective protein fractions, precipitate the serum or plasma protein from a known volume of serum or plasma using the standard trichloroacetic acid (TCA) precipitation method. Transfer the protein thus precipitated into scintillation fluid for further measurement as described above. Similar to the serum or plasma proteins, TCA precipitation of proteins of various tissues or organs may also be used to estimate the ^{14}C-R incorporation or accumulation in a specific tissue/organ. For urine analysis, the samples of urine collected by bladder puncture before harvesting other organs can be used for the presence of ^{14}C-R. Essentially, a known aliquot of the sample (50 μl to 100 μl) can be subjected to scintillation counting as described above.

3.2.5.3 Qualitative Analysis of Radioisotope

Autoradiography

Fluorography or autoradiography is a valuable analytical method that can provide a wealth of information on the level of alkylation or incorporation of radiolabeled, investigational drug [^{14}C-R]. While the TCA precipitation method (described above) will indicate the amount of radiolabel (^{14}C-R) incorporation into total serum or plasma protein, it does not provide details on the identity of protein species that are alkylated or targeted by ^{14}C-R. Gel autoradiography (one dimensional and two-dimensional approaches) combined with immunoblotting and/or LC-MS analysis will enable us to identify the nature of the protein(s) that are alkylated by ^{14}C-R.

Gel Autoradiography

In general, the plasma or serum is rich in proteins; hence it is always essential to determine the total protein concentration of plasma or serum samples. Also, in order to achieve proper resolution and optimal signal, it is important to know the quantity of total protein loaded onto the gels. Determine the protein concentration of samples. Perform SDS-PAGE analysis as per standard procedure or the supplier of gels, followed by colloidal Coomassie blue staining [Neuhoff et al. 1988] (Fig. 3.3). For

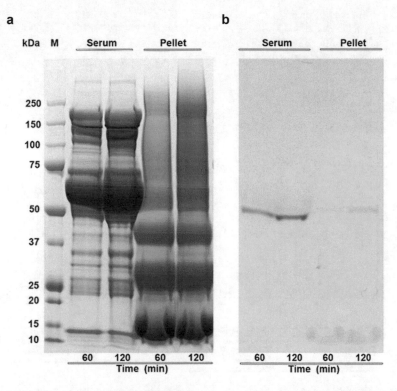

Fig. 3.3 Selective binding of ^{14}C-3-BrPA with rat serum proteins. (**a**) Coomassie stained SDS-PAGE gel showing the protein profile of serum and pellet of ^{14}C-3-BrPA dosed rat. (**b**) A corresponding autoradiogram showing time-dependent increase in autoradiogram signal in the serum, at ~50–60 kDa. Serum obtained from blood collected from the rats 60 min or 120 min after the 3-BrPA infusion indicate a time dependent increase in the amount of ^{14}C incorporation. [Reprinted by permission from BioMed Central Ltd. Springer Nature, BMC Res Notes, Systemic administration of 3-bromopyruvate reveals its interaction with serum proteins in a rat model, Kunjithapatham et al. Copyright © 2013]

2D-gel electrophoresis, prior to isoelectrofocusing (IEF), clean-up the samples using a 2D-Clean-up kit (GE-Healthcare). Perform IEF using Immobiline™ dry gel strips of linear pI (isoelectric point) range 3–10, 7 cm (GE-Healthcare) [Ganapathy-Kanniappan et al. 2009]. Following IEF, subject the gel strips to second-dimensional separation using Zoom gels as per supplier's instructions followed by Coomassie blue staining (Fig. 3.4a, b). If required, perform an additional step, incubate the gels with "Amplify" solution, to enhance the radioactive signal, before vacuum drying and exposure to X-ray film.

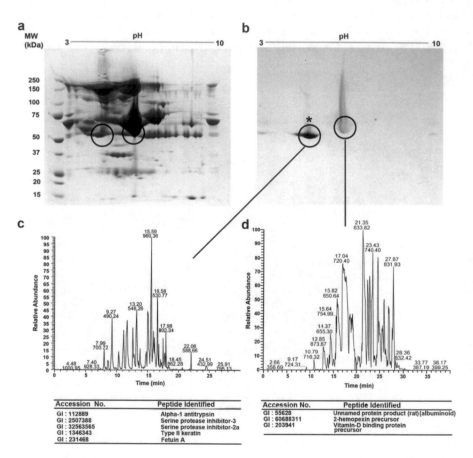

Accession No.	Peptide Identified
GI : 112889	Alpha-1 antitrypsin
GI : 2507388	Serine protease inhibitor-3
GI : 32563565	Serine protease inhibitor-2a
GI : 1346343	Type II keratin
GI : 231468	Fetuin A

Accession No.	Peptide Identified
GI : 55628	Unnamed protein product (rat)(albuminoid)
GI : 60688311	2-hemopexin precursor
GI : 203941	Vitamin-D binding protein precursor

Fig. 3.4 ¹⁴C-3-BrPA primarily binds with two peptides in rat serum. (**a**) A Coomassie stained 2D-gel showing serum protein spots after 120 min of 3-BrPA treatment and (**b**) a corresponding autoradiogram showing a strong and also a weak signal at ~50–60 kDa. (**c, d**) LC-MS/MS base-peak chromatograms of the peptide spots corresponding to strong (*) and weak signals on 2D-autoradiogram with the list of possible peptides, in the order of matched score. [Reprinted by permission from BioMed Central Ltd. Springer Nature, BMC Res Notes, Systemic administration of 3-bromopyruvate reveals its interaction with serum proteins in a rat model, Kunjithapatham et al. Copyright © 2013]

Immunoblotting and LC-MS/MS

To identify the protein spots corresponding to the radioactive signal, excise/extract, the gel spots from the dry-gel after autoradiography. Proteolyze the gel spots containing proteins with trypsin following standard procedure [Shevchenko et al. 1996], and perform LC-MS/MS identification of protein species (Figure 3.4c, d). Upon identification of possible proteins that correspond to radioactive signals on the auto-radiogram, verify the protein with immunoblotting. Perform a standard western blotting (immunoblotting) protocol using SDS-PAGE or 2D gels. Transfer the proteins resolved on the gel into PVDF membranes by conventional semi-dry or vacuum method followed by probing with specific antibodies. Adopt the protocol outlined by the specific antibody supplier's instructions for the use of primary antibodies and immunodetection.

Tissue Autoradiography

Use the formalin-fixed tissues and follow standard steps like dehydration by graded ethanol, embedding by Paraplast Plus wax (McCormick Scientific), sectioning at 5 μm, mounting on slides, oven drying, and deparaffinization. Subject the tissue sections to Hematoxylin and Eosin (H&E) staining [Ganapathy-Kanniappan et al. 2012] and view under a light microscope. To confirm apoptotic death, follow the staining procedure using commercially available apoptosis detection kits following the supplier's protocol.

For tissue-autoradiography, place the unstained slides derived from ^{14}C-R infused rat organs in a large X-ray cassette with intensifying screens, and expose them to X-ray film for 30 days at −80 °C. On the day of film development, remove the cassette from −80 °C, and allow it to come to room temperature (which might take a few hours) before developing the film. The autoradiogram can be compared with respective tissue sections stained with H & E to infer the presence or abundance of radioactive signal (Fig. 3.5).

3.2.6 Study Questions

(i) Why is the labeling (e.g., radioisotope) or tagging of a potential alkylating agent necessary to determine its pharmacokinetics?
(ii) What are the radioactive isotopes that may be employed to label a potential alkylating agent?
(iii) How does the half-life of a radioisotope impact the analysis?
(iv) How to differentiate the serum radioisotope estimate as a protein-bound or free, unbound radioisotope?

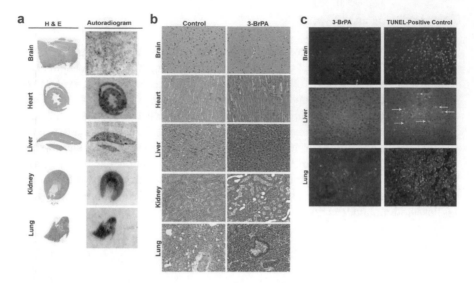

Fig. 3.5 Tissue distribution of ^{14}C-3-BrPA in systemic delivery. (**a**) Autoradiogram and corresponding H & E staining of tissue sections showing radioactive signal in organs such as liver, lung, kidney and heart but not brain. (**b**) H & E staining showing normal tissue architecture in control and 3-BrPA dosed rat tissues. (**c**) TUNEL-assay showing absence of positive staining in 3-BrPA dosed rat tissues indicating absence of apoptosis. For reference, TUNEL positive, control slides are included. [Reprinted by permission from BioMed Central Ltd. Springer Nature, BMC Res Notes, Systemic administration of 3-bromopyruvate reveals its interaction with serum proteins in a rat model, Kunjithapatham et al. Copyright © 2013]

3.3   Pharmacodynamics

3.3.1   Background

As discussed at the beginning of this chapter, the objective of pharmacodynamics (PD) is to assess the therapeutic response of a given drug. Since the PD evaluates the sensitivity and/or response of the disease (e.g., cancer) to a potential drug, the analytical methods rely on treatment-related alterations in specific cellular, molecular, biochemical, and functional indicators of the disease. In other words, PD enables the determination of desired phenotypic and/or pathophysiological changes that are attributable to potential therapeutic outcomes. Both in preclinical research and clinical investigation, multiple approaches are adopted to assess the therapeutic response to a potential drug. In general, they comprise non-invasive or minimally invasive imaging techniques (e.g., PET, CT, US, MRI), serum markers and/or biopsy to verify the histopathology. While some of the imaging modalities involve tracers like radioisotopes (e.g., ^{18}F in PET), the description of such clinical imaging techniques may be beyond the scope of the book. Hence, we will focus on techniques that are relevant in preclinical research and accessible to non-experts in the field.

Tracer techniques play a pivotal role in one of the components of PD, i.e., target validation also known as drug-specificity. Since the efficacy of a potential drug involves specific molecular targeting and the related mechanisms that initiate cascade(s) of events culminating in the desired phenotype, validation of target specificity is crucial in understanding the PD of a potential drug. Especially, if the potential agent or drug under investigation is in the preclinical phase, validation of target specificity in vivo is indispensable. In this context, if the drug under investigation is an alkylating agent or halogenated alkylating agent, tracer technique will enable us to determine its molecular target. For example, if the primary molecular target of the potential drug is a receptor or an enzyme, then radiolabeling the potential drug will enable us to validate whether the desired outcome/phenotype is achieved through the respective molecular (target) mechanism.

3.3.2 Method

In vivo evaluation of therapeutic efficacy of a drug (e.g., alkylating agent) requires an animal model that mimics human disease. In order to establish a preclinical disease model, the species selection, disease type, and method of disease initiation are critical. First and foremost, the best experimental organism in mammals that has been widely used in the laboratory is the mouse species, which has profoundly contributed to the advancement of our understanding of various diseases. In human cancer research, using an animal model, in particular, a murine model is essential. In general, to develop human cancer in a mouse species, the mouse needs to be immunocompromised (e.g., athymic nude mouse, SCID mouse) to evade immune rejection. Next, in cancer research, the type of tumor and method of tumor implantation may vary according to the investigator's objective. In general, human cancer cells (e.g., hepatocellular carcinoma, HCC) will be either injected into the right or left flank (subcutaneous tumor model) or as orthotopic implantation onto the respective organ (e.g., liver). While the former requires no surgery or incision, the latter requires survival surgery to expose the liver under anesthesia, injection of cells, followed by wound/incision closure, and recovery from anesthesia. Survival surgery is an intensive procedure, and detailed guidelines as outlined by the IACUC must be followed.

The effect of a potential therapeutic on a specific phenotype/disease is also used to validate the therapeutic target. It may be beneficial to the readers to understand some of the fundamental approaches in the evaluation of pharmacodynamics of a potential therapeutic in preclinical models. Here, using our experience in a mouse model of human cancer, we outline the strategy to develop a valid model that facilitates the assessment of tumor growth and response to therapy. As illustrated in the schematic (Fig. 3.6), initiation of the experiment involves the development of an ideal model and treatment with the radiolabeled agent/therapeutic under investigation. Next, the assessment of tumor response to therapy using imaging modality as well as histopathology. Finally, biochemical and functional analysis of the tumor to characterize the therapeutic target of the radiolabeled agent.

Fig. 3.6 Flow chart of experimental steps to identify the target of an alkylating agent or a potential therapeutic

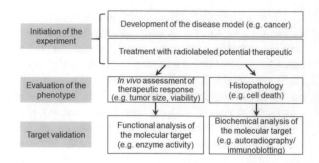

3.3.3 Principle

The principle underlying radiolabeling of the investigational, alkylating agent and their mechanism of alkylation are as described in Sect. 3.2.3.

3.3.4 Reagents/Equipment

In addition to all the reagents and chemicals and apparatus enlisted under Sect. 3.2.4, the following are required:

Radiolabeled compound (e.g., ^{14}C-3-BrPA), as in Sect. 3.2.3.
Experimental model (e.g., mouse model of human cancer).
Electrophoretic chemicals and reagents as described elsewhere.
Chemicals/stains specific for histopathology and histochemistry.
Autoradiography requirements such as X-ray film, and film processor/developer.
Tissue homogenizer.
Electrophoretic apparatus and powerpack.
Microcentrifuge, table-top refrigerated centrifuge.

3.3.5 Stepwise Protocol

3.3.5.1 Imageable Tumor Model

Prior to the discussion on the particular tracer technique for target validation, it is essential to adopt a relevant and feasible method for the assessment of therapeutic response or phenotypic alteration. In other words, in the case of preclinical models of human cancer (e.g., mouse), in vivo imaging, also known as the live-animal imaging of tumor's response to treatment is a widely adopted protocol. The advantage of such in vivo imaging is that it will enable non-invasive evaluation of the outcome of therapy before euthanasia. However, in general, such imaging modalities require a reporter that will distinguish the disease (e.g., cancer) in the host (mouse) background. Among various types of reporters, fluorescence- and bioluminescence-based indicators are more commonly used due to the ease of handling in imaging. Hence, in a preclinical model of human cancer, the carcinogenic cells are engineered to harbor a reporter of choice to serve as an indicator of tumor growth and viability. The commonly used reporters are luciferase (*luc*) reporter (for bioluminescence imaging) and fluorescent reporter (e.g., green fluorescent protein, GFP) (for fluorescent imaging) among others. The *luc*-reporter regulates the expression of the enzyme, luciferase which acts upon the substrate luciferin, to release the luminescence signal. Typically, cancer cells expressing the luciferase enzyme will be implanted onto the animal model, and following the initial period of tumor growth, the substrate luciferin will be injected into the animal model via the tail-vein or intraperitoneal injection. Since the luciferase expression is selectively confined to the tumor growth, the enzymatic reaction on the substrate (luciferin) at the tumor site will emit a bright luminescence signal, indicating tumor growth, size, etc. The signal is then captured by the imaging system through specific equipment (e.g., IVIS spectrum in vivo imaging system, Perkin Elmer Co., MA, USA). Since luciferase is seldom present in rodents or humans, bioluminescence imaging (BLI) (signal intensity) will indicate tumor growth and viability. If an investigative drug has potent anticancer effects, it will block the progression of tumor growth which will be reflected on the BLI signal intensity. An example of such imaging to evaluate tumor growth and response to therapy is illustrated in Fig. 3.7. Similarly, other reporters may also be used. In conditions where cancer cells were used without any reporter, the physical appearance of tumor growth and size may be determined based on palpability and caliper measurements, particularly in subcutaneous tumor models.

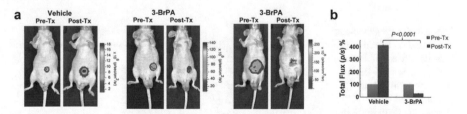

Fig. 3.7 Percutaneous injection of 3-BrPA affects *luc*-Hep3B tumor in mice. (**a**) *Luc*-Hep3B tumor implantation and 3-BrPA (1.75 mmol/l) treatment were followed by bioluminescence imaging of representative mice before *(Pre-Tx)* and after *(Post-Tx)* treatment. (**b**) Graph shows mean quantification of bioluminescence imaging signal, with a significant decrease in intensity in 3-BrPA–treated mice ($n = 6$). *P* value is from two-sample *t* test comparing groups after treatment. Error bars = standard error of the mean. p = photons, p/s = photons per second, sr = steradian. 3-BrPA- 3-bromopyruvic acid. [Reproduced with permission from Radiology (Ganapathy-Kanniappan et al., 2012), Radiological Society of North America (RSNA) publishers]

3.3.5.2   Target Validation

In tracer technique-based target validation, the principal requirement is the radio-labeled compound (i.e., the alkylating agent under investigation). Here, we will use the example as outlined in Sect. 3.2.3. Next, the methods employed for target validation may include functional and/or biochemical assays pertinent to the molecular target. (i) Functional assay: If the potential alkylating agent's primary target is an enzyme, upon alkylation the enzyme's structure and/or function may be altered irreversibly. Thus, analysis of the specific enzyme's activity may indicate the alkylation-dependent loss of function. For example, in cancer cells the primary target of the alkylating agent, 3-BrPA is the enzyme glyceraldehyde-3-phosphate dehydrogenase (GAPDH). Hence upon treatment with 3-BrPA tumor-GAPDH activity is severely impaired due to the specific alkylation by the alkylating agent (Fig. 3.8). Similarly, if the molecular target is a receptor that is alkylated by the potential alkylating agent, analysis of the receptor function (i.e., ligand binding efficiency) will verify treatment-dependent functional impairment of the target. Thus, functional analysis of the molecular target (e.g., enzyme, receptor) will validate the potential alkylating agent's specificity and the related mechanism underlying the phenotypic effects.

Next, biochemical analysis of the molecular target may be performed by several approaches, and here we will discuss principal methods that rely on the tracer (radioisotope) used to label the alkylating agent. For instance, gel-autoradiography followed by target verification by LC-MS/MS verification or immunoblotting (Fig. 3.9).

Next, immunoprecipitation of the potential molecular target followed by autoradiography to investigate the alkylation of the target (Fig. 3.10). While other methods may be relevant, the protocols described here are the common and widely employed techniques relevant in radioisotope-labeled alkylating agents.

Fig. 3.8 Graph shows that percutaneous injection of 3-BrPA affects GAPDH activity ($n = 6$). 3-BrPA-3-bromopyruvic acid. [Reproduced with permission from Radiology (Ganapathy-Kanniappan et al., 2012), Radiological Society of North America (RSNA) publishers]

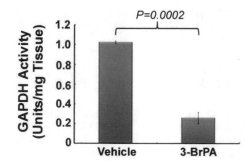

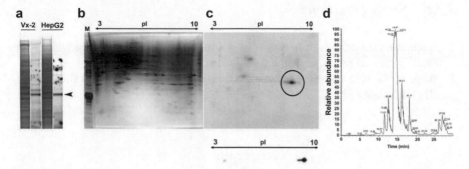

Fig. 3.9 Glyceraldehyde-3-phosphate dehydrogenase (GAPDH) pyruvylation upon 3-bromopy-ruvate treatment. (**a**) Coomassie blue-stained gels and their corresponding autoradiograms are shown for HepG2 and Vx-2 cell lines treated with ^{14}C-3-bromopyruvate. The arrowhead indicates the gel band excised and subjected to mass spectrometric characterization. (**b**) Silver-stained 2D gel of whole cell lysate from SK-Hep1 cells treated with ^{14}C-3-bromopyruvate and its corresponding autoradiogram (**c**) showing ^{14}C-3-bromopyruvate incorporation. The circle indicates the intense gel spot excised and characterized by mass spectrometry. *Bottom panel*: Immunoblot of a 2D gel showing GAPDH as the spot identified on 2D autoradiogram. (**d**). Liquid chromatographic-mass spectrometry chromatogram showing GAPDH to be the peptide identified from autoradiogram spot of the 2D gel indicated by circle (in **c**). [Reproduced with permission from Anticancer Research (Ganapathy-Kanniappan et al., 2009), IIAR publishers]

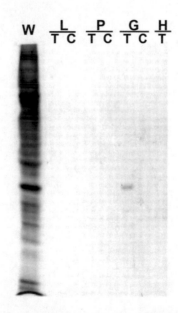

Fig. 3.10 Autoradiogram of immunoprecipitates from SK-Hep1 cells treated with ^{14}C-3-bromopyruvate, showing the ^{14}C incorporation. Lanes: W, whole cell lysate; lanes L, P, G, and H correspond to reactions to specific antibodies against the targets lactate dehydrogenase (L), pyruvate dehydrogenase (P), glyceraldehyde-3-phosphate dehydrogenase (G) and hexokinase type II (H), respectively. T, Target-specific antibody, and C, Control IgG for the respective target. [Reproduced with permission from Anticancer Research (Ganapathy-Kanniappan et al., 2009), IIAR publishers]

3.3.6   Study Questions

1. What are the methods of noninvasive animal imaging that are employed in the investigation of preclinical models of human disease?
2. What is the role of radioisotopes in target validation of potential therapeutic, alkylating agents?

References

Ganapathy-Kanniappan, S., Geschwind, J. F., Kunjithapatham, R., Buijs, M., Vossen, J. A., Tchernyshyov, I., Cole, R. N., Syed, L. H., Rao, P. P., Ota, S., & Vali, M. (2009). Glyceraldehyde-3-phosphate dehydrogenase (GAPDH) is pyruvylated during 3-bromopyruvate mediated cancer cell death. *Anticancer Research, 29*(12), 4909–4918.

Ganapathy-Kanniappan, S., Kunjithapatham, R., Torbenson, M. S., Rao, P. P., Carson, K. A., Buijs, M., Vali, M., & Geschwind, J. F. (2012). Human hepatocellular carcinoma in a mouse model: Assessment of tumor response to percutaneous ablation by using glyceraldehyde-3-phosphate dehydrogenase antagonists. *Radiology, 262*(3), 834–845.

*Kunjithapatham, R., Geschwind, J. F., Rao, P. P., Boronina, T. N., Cole, R. N., & Ganapathy-Kanniappan, S. (2013). Systemic administration of 3-bromopyruvate reveals its interaction with serum proteins in a rat model. *BMC Research Notes, 6*, 277.

Neuhoff, V., Arold, N., Taube, D., & Ehrhardt, W. (1988). Improved staining of proteins in polyacrylamide gels including isoelectric focusing gels with clear background at nanogram sensitivity using Coomassie brilliant blue G-250 and R-250. *Electrophoresis, 9*(6), 255–262.

Roberts, S. A. (2003). Drug metabolism and pharmacokinetics in drug discovery. *Current Opinion in Drug Discovery & Development, 6*(1), 66–80.

Shevchenko, A., Wilm, M., Vorm, O., & Mann, M. (1996). Mass spectrometric sequencing of proteins silver-stained polyacrylamide gels. *Analytical Chemistry, 68*(5), 850–858.

Further Reading

Amanchy, R., Kalume, D. E., & Pandey, A. (2005). Stable isotope labeling with amino acids in cell culture (SILAC) for studying dynamics of protein abundance and posttranslational modifications. *Science's STKE, 2005*(267), pl2.

Appel, M., Couderc, E., & Feger, J. (1992). Comparative studies of protein biosynthesis: The main experimental parameters in pulse and pulse-chase experiments must be standardized. *Biology of the Cell, 74*(2), 235–238.

Boffey, S. A. (1987). Autoradiography and fluorography. In J. M. Walker & W. Gaastra (Eds.), *Techniques in molecular biology* (p. 288). Springer.

Bonifacino, J. S. (2001). Metabolic labeling with amino acids. *Current Protocols in Protein Science*. Chapter 3:Unit 3.7.

© Springer Nature Switzerland AG 2022   61
S. Ganapathy-Kanniappan, *Isotopic Tracer Techniques in Preclinical Research*,
Techniques in Life Science and Biomedicine for the Non-Expert,
https://doi.org/10.1007/978-3-030-99700-7

Bosch, F., & Rosich, L. (2008). The contributions of Paul Ehrlich to pharmacology: A tribute on the occasion of the centenary of his Nobel prize. *Pharmacology, 82*(3), 171–179.

Classic Protocol. (2005). Northern blotting: Transfer of denatured RNA to membranes. *Nature Methods, 2*(12), 997–998.

den Broeck, V., & Walter, M. M. (2015). Chapter 3—Drug targets, target identification, validation, and screening. In C. G. Wermuth, D. Aldous, P. Raboisson, et al. (Eds.), *The practice of medicinal chemistry* (4th ed., p. 45). Academic Press.

Elagib, K. E., Xiao, M., Hussaini, I. M., Delehanty, L. L., Palmer, L. A., Racke, F. K., Birrer, M. J., Ganapathy-Kanniappan, S., McDevitt, M. A., & Goldfarb, A. N. (2004). Jun blockade of erythropoiesis: Role for repression of GATA-1 by HERP2. *Molecular and Cellular Biology, 24*(17), 7779–7794.

Fan, J., & de Lannoy, I. A. (2014). Pharmacokinetics. *Biochemical Pharmacology, 87*(1), 93–120.

Farrell, J., & Robert, E. (2010). *RNA methodologies A laboratory guide for isolation and characterization* (4th ed.). Elsevier.

Fritzsche, S., & Springer, S. (2014). Pulse-chase analysis for studying protein synthesis and maturation. *Current Protocols in Protein Science, 78*, 30.3.1–30.3.23.

Gallagher, P. E., & Diz, D. I. (2001). Analysis of RNA by northern-blot hybridization. *Methods in Molecular Medicine, 51*, 205–213.

Ganapathy-Kanniappan, S., Geschwind, J. F., Kunjithapatham, R., Buijs, M., Syed, L. H., Rao, P. P., Ota, S., Kwak, B. K., Loffroy, R., & Vali, M. (2010). 3-bromopyruvate induces endoplasmic reticulum stress, overcomes autophagy and causes apoptosis in human HCC cell lines. *Anticancer Research, 30*(3), 923–935.

Hellman, L. M., & Fried, M. G. (2007). Electrophoretic mobility shift assay (EMSA) for detecting protein-nucleic acid interactions. *Nature Protocols, 2*(8), 1849–1861.

Hutchinson, O. C., Collingridge, D. R., Barthel, H., Price, P. M., & Aboagye, E. O. (2003). Pharmacodynamics of radiolabelled anticancer drugs for positron emission tomography. *Current Pharmaceutical Design, 9*(11), 931–944.

Jager, P. L., Vaalburg, W., Pruim, J., de Vries, E. G., Langen, K. J., & Piers, D. A. (2001). Radiolabeled amino acids: Basic aspects and clinical applications in oncology. *Journal of Nuclear Medicine, 42*(3), 432–445.

Kim, I. Y., Suh, S. H., Lee, I. K., & Wolfe, R. R. (2016). Applications of stable, nonradioactive isotope tracers in vivo human metabolic research. *Experimental & Molecular Medicine, 48*, e203.

Krebs, N. E., & Hambidge, K. M. (2001). Zinc metabolism and homeostasis: The application of tracer techniques to human zinc physiology. *Biometals, 14*(3–4), 397–412.

McCabe, B. J., & Previs, S. F. (2004). Using isotope tracers to study metabolism: Application in mouse models. *Metabolic Engineering, 6*(1), 25–35.

Quemeneur, E. (2000). Autoradiography and fluorography. In R. Rapley (Ed.), *The nucleic acid protocols handbook* (p. 169). © Humana Press.

Schwab, D., Portron, A., Backholer, Z., Lausecker, B., & Kawashima, K. (2013). A novel double-tracer technique to characterize absorption, distribution, metabolism and excretion (ADME) of [14C]tofogliflozin after oral administration and concomitant intravenous microdose administration of [13C]tofogliflozin in humans. *Clinical Pharmacokinetics, 52*(6), 463–473.

Simon, E., & Kornitzer, D. (2014). Pulse-chase analysis to measure protein degradation. *Methods in Enzymology, 536*, 65–75.

Southern, E. (2006). Southern blotting. *Nature Protocols, 1*(2), 518–525.

Southern, E. M. (1975). Detection of specific sequences among DNA fragments separated by gel electrophoresis. *Journal of Molecular Biology, 98*(3), 503–517.

Spanos, C., & Moore, J. B. (2016). Sample preparation approaches for iTRAQ labeling and quantitative proteomic analyses in systems biology. *Methods in Molecular Biology, 1394*, 15–24.

Strauss, H. W., & Pitt, B. (1978). Evaluation of cardiac function and structure with radioactive tracer techniques. *Circulation, 57*(4), 645–654.

Streit, S., Michalski, C. W., Erkan, M., Kleeff, J., & Friess, H. (2009). Northern blot analysis for detection and quantification of RNA in pancreatic cancer cells and tissues. *Nature Protocols, 4*(1), 37–43.

Sun, R. C., Fan, T. W., Deng, P., Higashi, R. M., Lane, A. N., Le, A. T., Scott, T. L., Sun, Q., Warmoes, M. O., & Yang, Y. (2017). Noninvasive liquid diet delivery of stable isotopes into mouse models for deep metabolic network tracing. *Nature Communications, 8*(1), 1646-z.

Unwin, R. D., Griffiths, J. R., & Whetton, A. D. (2010). Simultaneous analysis of relative protein expression levels across multiple samples using iTRAQ isobaric tags with 2D nano LC-MS/MS. *Nature Protocols, 5*(9), 1574–1582.

Vértes, A., Nagy, S., Klencsár, Z., Lovas, R. G., & Rösch, F. (Eds.). (2011). *Handbook of nuclear chemistry* (2nd ed.). Springer.

Wagenmakers, A. J. (1999). Tracers to investigate protein and amino acid metabolism in human subjects. *The Proceedings of the Nutrition Society, 58*(4), 987–1000.

Index

© Springer Nature Switzerland AG 2022
S. Ganapathy-Kanniappan, *Isotopic Tracer Techniques in Preclinical Research*,
Techniques in Life Science and Biomedicine for the Non-Expert,
https://doi.org/10.1007/978-3-030-99700-7

Printed in the United States
by Baker & Taylor Publisher Services